H2Optimization

Dr. Barry L. Liner, P.E.

Dr. Holger R. Maier

H2Optimization: An Introduction to Optimization and Operations Research for Infrastructure Professionals

www.h2optimization.com

Aldera

Published in collaboration with

Water Anthology Press

www.WaterAnthology.com

Cover art provided by: SelfPubBookCovers.com/RLSather

ISBN: 0996580409
ISBN-13: 978-0-9965804-0-3

DEDICATION

To all water sector professionals who work every day to protect public health and drive economic development through their dedicated efforts to provide sustainable solutions for management of the world's most precious resource, water.

CONTENTS

ACKNOWLEDGMENTS

A number of people have been influential in helping develop this book including Britt Sheinbaum, Donna Vincent Roa, David McIver, Corey Williams, Jill Means, Kevin Shipp, Jeff Frey, Nick Gardner, David Binning, Sharon deMonsabert, Mark Houck, Seth Brown, Kevin Morley, Pam Kenel, Anthony Damiani, and all of the students I had the pleasure of teaching during my years at George Mason University.

PREFACE

Many times, when people hear the term "optimization," they think of geeky mathematical gurus, often with titles like management scientist, operations researcher, or systems engineer. We know a civil engineer builds things like water and wastewater facilities; a mechanical engineer works with things like Heating, Ventilation and Air Conditioning (HVAC); and an electrical engineer works with circuit boards or power plants. But an industrial or systems engineer? What do they do and how can they know about our problems? When talking about optimization, these systems engineers simply plug a bunch of numbers into a black box and an answer comes out that they say is the best solution. Why should anyone trust these results? Who knows if this black box suffered from the problem of garbage in- garbage out (GIGO)? Who is to say that this approach isn't just the recommendations of the proverbial snake oil salesperson?

Let's take a look at the industrial engineering, or systems engineering, profession. According to the Institute of Industrial Engineers (IIE):

> *Industrial Engineering is concerned with the design, improvement, and installation of integrated systems of people, materials, information, equipment, and energy. It draws upon specialized knowledge and skill in the mathematical, physical, and social sciences, together with the principles and methods of engineering analysis and design, to specify, predict and evaluate the results obtained from such systems.*

The International Council on Systems Engineering (INCOSE) defines systems engineering as such:

> *Systems Engineering is an interdisciplinary approach and means to enable the realization of successful systems. It focuses on defining customer needs and required functionality early in the development cycle, documenting requirements, then proceeding with design synthesis and system validation while considering the complete problem Systems engineering integrates all the disciplines and specialty groups into a team effort forming a structured development process that proceeds from concept to production to operation. Systems engineering considers both the business and the technical needs of all customers with the*

goal of providing a quality product that meets the user needs.

Most systems engineers accept the following basic core concepts:

- *Understand the whole problem before you try to solve it.*
- *Translate the problem into measurable requirements.*
- *Examine all feasible alternatives before selecting a solution.*
- *Make sure you consider the total system life cycle. The birth to death concept extends to maintenance, replacement and decommission. If these are not considered in the other tasks, major life cycle costs can be ignored.*
- *Make sure to test the total system before delivering it.*
- *Document everything.*

A systems engineer seeks to integrate all the viewpoints of a problem and tries to provide the best solution according to a rather rigorous problem-solving approach. That sounds pretty good. Now that we're comfortable with the systems engineer, what about the optimization black box?

The techniques used in the black box "optimizer" generally do involve calculus and linear algebra, and that scares people who aren't comfortable with higher-level mathematics. However, while the engine that runs the algorithms to solve the problem uses significant computing power, the problem definition consists of a set of basic principles that every decision maker should know, whether using optimization or not.

The purpose of this book is to provide some insight into the definition of problems that can then take advantage of the computing power of optimization to provide the most efficient and effective solution to a problem. It is critical that decision makers fully define what problem they want to solve so they can help control the quality of the problem definition, which will help ensure that the solution provided is not a result of GIGO, but rather a better option than would be found using traditional techniques. If you want to examine these techniques applied in a similar context with more depth, we recommend <u>Systems Analysis for Sustainable Engineering: Theory and Applications</u> by Ni-Bin Chang (McGraw Hill Professional, 2010, ISBN 0071630066, 9780071630061).

In addition, this book is focused on the water infrastructure sector, including drinking water, wastewater, stormwater, and even solid waste and energy. Many other sectors use optimization techniques, such as airlines that use logistics to optimize flight routes, staff allocation, and facility location. With the water sector facing billions of dollars of need to address aging infrastructure, optimization can help reduce capital and operations and maintenance (O&M) costs, which can help the sector face those huge infrastructure challenges while lessening the impact on ratepayers and the public sector coffers.

HOW TO USE THIS BOOK

This book is divided into three parts to provide the most beneficial use for a variety of readers.

Part I: Structure of Optimization Problems (Chapters 1 and 2) delivers an overview of how optimization problems can be set up and provides examples. The goal is to familiarize the reader with the overall process of optimization. If you can set up a simple problem with three choices, then you can set up a larger problem with hundreds or thousands of choices using the same approach, the only difference being the computational power needed to solve the problem.

Part II: Inside the Black Box (Chapters 3 through 5) introduces optimization algorithms for those readers looking to understand how the problem solving engines work. The discussion will focus primarily on linear programming and genetic algorithms. We will not get into the detailed mathematics behind the algorithms, but rather seek to provide a comfort level for the reader about the operation of the analytical engines that provide the solutions.

Part III: Food For Thought (Chapters 6 through 8) utilizes research case studies to provide examples of how optimization can be applied to water resources.

References

Institute of Industrial Engineers (IIE) website:
http://www.iienet2.org/Details.aspx?id=2644

International Council on Systems Engineering (INCOSE) website:
http://www.incose.org/practice/whatissystemseng.aspx

PART I: STRUCTURE OF OPTIMIZATION PROBLEMS

1 INTRODUCTION TO OPTIMIZATION

Before we begin, let us remember that **Optimization as a tool does not replace the role of civil engineering or CFO. In fact, it makes them more valuable. Who better to set up problems to ensure that the optimization algorithms meet their potential than knowledgeable engineers and controllers?**

So what is **optimization**? According to dictionary.com, optimize is a verb with the following meanings:

1. to make as effective, perfect, or useful as possible.
2. to make the best of.
3. *Computers.* to write or rewrite (the instructions in a program) so as to maximize efficiency and speed in retrieval, storage, or execution.
4. *Mathematics.* to determine the maximum or minimum values of (a specified function that is subject to certain constraints).

When systems engineers speak of optimization, they are generally talking about the fourth definition. However, these engineers often also use technical vernacular that can obfuscate the functional objective by causing the operational, fiscal, and capital constraints to be inaccurately defined. For example, boolean and non-linear relationships, such as quadratic or discrete integer constraints, may be oversimplified into linear, continuous equations that may be unrealistic, or could lead to poor assumptions of the feasibility of the solution set. (Exactly!)

In reality, we should be focusing on the first two definitions. Given budgetary limitations and facing large infrastructure needs, decisions need to be made to spend our money in the most effective or useful means possible. That is, let's make the best of the situation. (Much better!)

Problem Definition

To discuss optimization problems, this book will use the Linear Programming (LP) framework. LP will be discussed later, but we'll focus on problem definition first. Problems that are conducive to using optimization techniques generally have the following characteristics:

- *A well-defined objective*. An example might be: To meet the infrastructure demands of a growing population at the lowest cost possible
- *Courses of action*. Examples include: Different sets of pipes and storage that could potentially be constructed or rehabilitated; or different types of financing mechanisms (bonds, rates, public private partnership, alternative service delivery, etc.)
- Total achievement of the objective must be *constrained by scarce resources*. These can be a budget limit or physical limits such as the fact that you can't put 100 cubic meters of water through a pipe that can only handle 20.

These characteristics sound like many problems we face, right? If a problem has these characteristics, it can usually be addressed by linear programming or similar optimization techniques. The steps to define the problem that can then be translated into the equations needed by the mathematical engine are pretty straightforward, as shown in Figure 1.

Using Optimization to Solve Problems

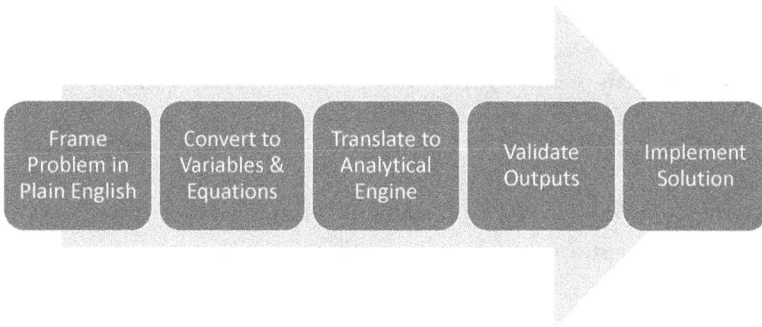

Figure 1 General optimization solution framework.

Frame the problem using plain English

- Define what you want to accomplish. This is the objective function.
- Identify the things that can be changed, that is, decisions about whether or not to do something, or how much of something to use. These are decision variables.
- Identify the limiting factors, or constraints. Types of constraints could include budgets, permit compliance, capacity, and the like.

Convert components to variables and equations

- For each decision variable, determine the type of decision. These could include:
    - Yes/no, or binary variable like *build a tank* or *not build a tank*.
    - Linear quantity, like how large to build a tank.
    - Round number or integer quantity like how many staff (you have to have 7 or 8 people, not 7.39 people).
    - Multiple mutually exclusive choices like do nothing, rehabilitate the pipe, or replace the pipe.
- For the objective function, specifically state whether a maximization or minimization of a specific equation is necessary. The objective function is generally minimizing cost or maximizing

profit, but as we will discuss later, there are a number of ways to bring in other concepts.

- Identify the constraints and write expressions in terms of both the variables and the limitations. For example, the flow available to a specific pressure zone in a city's distribution system must be greater than or equal to 100 cubic meters per day, or the impervious area of a development must be less than 10 percent of the total land area.

Translate the variables and equations into the format required by the analytical engine and run the optimization program.

- Let the analytical engine do the work of enumerating and evaluating the alternatives.

Validate the results (outputs) from the analytical engine

- Ensure that the suggested plan of action (set of decision variables) is truly feasible. This may be done by plugging the data back into a utility's hydraulic model, geographic information system (GIS), or financial models.
- Perform sensitivity analysis by looking at the optimal result and some near-optimal solutions. For example, perhaps the top 10 best solution sets can be evaluated for final due diligence for factors that might not have been directly calculated in the optimization program, such as the number of jobs created by a certain project.

Implement the solution and re-evaluate as new data become available.

- This can be done at the planning level, by running an optimization of capital projects on an annual basis. It can also be done on a real-time basis by evaluating which pumps and tanks to turn on and off for operations.

Again, optimization as a tool does not replace the role of civil engineering or CFO. In fact, it makes them more valuable. Who better to set up problems to ensure that the optimization algorithms meet their potential than knowledgeable engineers and controllers?

Let's apply this approach to a problem that appears complex, but actually turns out to be relatively straightforward.

Deconstructing the Problem

Suppose there are three industrial plants discharging into the Shenandoah River: Poultry Co., Chip Factory, and Dominion Chemical. Each plant discharges wastewater with Phosphorus and Nitrogen wastes. In order to improve water quality and meet the TMDL (Total Maximum Daily Load) in the river, the pollution must be reduced. It costs $15 to process a ton of Poultry Co. waste, which reduces the Phosphorus pollution by 200 pounds and reduces Nitrogen by 900 pounds. At the Chip Factory, $10 per ton treats waste to reduce Phosphorus by 400 pounds and Nitrogen by 500 pounds. At Dominion Chemical, reduction of 800 pounds of Phosphorus and 600 pounds of Nitrogen can be obtained by treating 1 ton of waste, at a cost of $20 per ton. All three plants are owned by the same holding company, who wants to minimize the costs of treatment while meeting the DEQ (Department of Environmental Quality) requirements of reducing Phosphorus pollution by 30 tons and Nitrogen loads by 40 tons in the river.

Let's take a look at a potentially equitable approach. If each facility was required to remove at least a third of the pollutants, the treatment requirements at each facility would be:

- Poultry Co. = 100 tons
- Chip Factory = 53.3 tons
- Dominion Chemical = 44.4 tons

This solution would yield a cost of over $2922. Now let's see what we can do with optimization.

Frame the Problem in Plain English

The objective is to meet the pollution reduction goals at minimum cost to the holding company.

The decision variables are how much treatment is needed at each plant, as the decision maker can control how much to treat at each plant.

The limiting factors are that the company must meet regulatory requirements by reducing Phosphorus by 30 tons and Nitrogen by 40 tons.

Convert components to variables and equations

The decision variables are the amount of tons processed at Poultry Co., Chip Factory, and Dominion Chemical. Let's set the following variables:

- P = number of tons processed at Poultry Co.
- C = number of tons processed at Chip Factory
- D = number of tons processed at Dominion Chemical

The objective function is to minimize total cost. Total cost is calculated by multiplying the unit costs times the number of tons treated at each location.

- Minimize Total Cost = 15 P + 10 C + 20 D

There are two constraints: Phosphorus reduction and Nitrogen reduction. The amount of each of the pollutants removed varies by facility, with the unit removal rates specified in the problem.

- Phosphorus removal constraint = 200 pounds of Phosphorus will be removed for each ton of waste processed at Poultry Co. Similarly, at Chip Factory and Dominion Chemical, each ton of wastewater treated removes 400 and 800 pounds of Phosphorus, respectively. The total of these must be greater than or equal to 30 tons (60,000 pounds) in order to meet the TMDL regulation. Thus the constraint can be written as:
  - o 200 P + 400 C + 800 D ≥ 60,000
- Following the same approach for the other pollutant makes the equation for the Nitrogen constraint:
  - o 900 P + 500 C + 600 D ≥ 80,000

Translate to format for analytical engine

Minimize Total Cost = 15 P + 10 C + 20 D

s.t. (subject to the constraints)

Phosphorus: 200 P + 400 C + 800 D ≥ 60,000

Nitrogen: 900 P + 500 C + 600 D ≥ 80,000

When this model is run by an analytical engine, the Minimum cost is $1577 with the following number of tons treated at each facility:

- Poultry Co. = 7.7 tons
- Chip Factory = 146.2 tons
- Dominion Chemical = 0 tons

This means that the company could save 46% by focusing on a small treatment facility at the Poultry Co. and a large facility at the Chip Factory, while the Dominion Chemical facility doesn't require any additional treatment.

Validate the results (outputs) from the analytical engine

To paraphrase a famous quote from George E. P. Box, "all models are incomplete, but some are useful." A well-constructed optimization model is a very useful mathematical model, but can never fully incorporate every nuance in a situation. This is where the solution validation by the decision maker comes in. The informed decision maker knows that the optimized solution saves the overall company 46%, but how does that impact each of the three plants? Does the Chip Factory manager feel like they got singled out? Does the Dominion Chemical manager feel that they got a free pass? What type of accounting framework needs to go into effect to recognize the overall savings without negatively impacting the Chip Factory, whose treatment costs will go up?

Again, as we stated at the beginning of this chapter, ***Optimization as a tool does not replace the role of civil engineering or CFO. In fact, it makes them more valuable. Who better to set up problems to ensure that the optimization algorithms meet their potential than knowledgeable engineers and controllers?***

2 OPTIMIZATION PROBLEMS

Many real world problems can be approximated by optimization problems. There are well-known successful applications in

- Logistics
- Manufacturing
- Marketing
- Finance and Investment
- Advertising
- Agriculture

To provide examples on how to set up optimization problems, we will use Linear Programming (LP) as a general framework, but the techniques are relevant to all optimization methods. As we have seen from the examples so far, linear programming can take a problem and put it in a format that allows an analytical engine to solve the problem. Linear programming models have the following assumptions:

- The parameter values are known with certainty (no randomness).
- The objective function and constraints exhibit constant returns to scale (linear)
- There are no interactions between the decision variables (assumption of additivity).
- Variables can take on any value within a given feasible range (continuity).

Linear programming is a good way to illustrate the potential of optimization techniques. Linear programs are deterministic (no uncertainty), but there are ways to address uncertainty through probabilistic or stochastic means. You can even "trick" linear programs by using techniques such as chance-constrained optimization, but these techniques are outside the scope of this book.

The intent of the first part of this book is for the reader to understand the general mechanics of the development and solution of problems, not to focus on the inner workings of the black box – algorithms which will be discussed in Part II (Chapters 3 through 5). Simple problems like the ones in this book, can be solved in MS Excel using add-ins such as the Solver or What's Best? There are even mobile apps to handle small linear programs. More robust solvers and programming languages can handle larger problems, such as how to route an airline or make staffing assignments to meet overnight delivery of packages. Some of these can be run on personal computers and some require networks of servers in the cloud to crunch the numbers. The output generated from optimization efforts can provide useful "what if" analysis.

Problem Setup

This chapter presents some solved linear programs to see how you can set up problems that might arise in the infrastructure sectors. These examples are simplified with fewer variables for demonstration and can all be easily solved using spreadsheets with add-ins. Of course, much larger problems (hundreds or thousands of variables) can be solved using more robust solvers and cloud computing. There are dozens of software packages available including general purpose optimization software such as *AMPL* and *GAMS*, task specific solvers such as *OptaPlanner* for resource allocation, and industry sector specific software such as that provided by companies such as *Optimatics* in the water sector. (NOTE: Product names, logos, brands and other trademarks referred to within this text are the property of their respective trademark holders. These are listed to demonstrate concepts only and the authors do not make any claims to the efficacy of the products mentioned.)

Regardless of the software package, remember that we need three key things when setting up an optimization problem:

- A well-defined objective.
- Alternative courses of action to choose from.
- Scarce resources that constrain the objective.

Optimization objectives generally fall into the categories of maximizing profit (or production) or minimizing cost (or negative impact). There are more advanced techniques that can handle multiple competing objectives (an example is presented in Part III) but the examples in Part I will only have a single objective.

Constraints

Constraints, on the other hand, come in many flavors. Common types of constraints are listed below:

- **Upper and lower bounds**. Examples include a maximum volume of 20 cubic meters $(V \leq 20)$ or a minimum of 10 cubic meters of water to meet demand in some period $(V \geq 10)$. Non-negativity constraints (≥ 0) are probably the most prevalent examples of lower bounds.
- **Limited resources**. For example, if you have a 40-hour workweek and it takes 4 hours to do task A, 6 hours to do task B and 7 hours to do task C, the constraint would be written as $4A + 6B + 7C \leq 40$.
- **Coverage constraints (requirements)**. If 1000 gallons of water must be captured and each rain barrel (RB) captures 55 gal and each cistern (C) captures 400 gal, then the constraint would be 55 RB + 400 C $\geq$ 1000.
- **Relative constraints (ratio)**. If you must have four times as many chairs as you do tables, then the constraint would be written *Chairs* $\geq$ 4* *Tables* or, alternatively, *Chairs* $-$ 4 **Tables* $\geq$ 0. In linear programming, you write the equation this way, because using a ratio of *Chairs/Tables* $\geq$ 4 is nonlinear because we are dividing two variables.
- **Relative constraints (percentage)**. Consider that a Water Resource Recovery Facility (WRRF) makes three types of water, ultrapure (U), agricultural irrigation (A), and groundwater recharge (G), and the WRRF must recharge 40% of the total water produced. Therefore, the constraint would be written as $0.6G$ -

$0.4U$ -$0.4A$ >0. The derivation of this linear constraint from its original non-linear representation is:

- $G \geq 0.4 *(G+U+A)$ rearranges to
- $G/(G+U+A) \geq 0.4$, which can be distributed to
- $G \geq 0.4G +0.4U + 0.4A$, which simplifies to
- $0.6G$ -$0.4U$ -$0.4A \geq 0$

- **Conservation or balance.** For example, the volume of a reservoir at the end of March (**VM**) is the volume at the end of February (**VF**) plus the inflows due to precipitation in March (**PM**) minus the outflows from dam releases during the month of March (**OM**). Thus, the constraint would be $VM = VF + PM - OM$.

- **Indicator variables.** Yes/no, or binary variable like *Build a tank* or *Not build a tank*. Technically, binary or integer variables require the analytical technique integer programming or mixed integer programming, but for the purposes of this text, when we refer to LP generally, we are including related techniques like integer programming, as well.

There are other constraints, of course, but these are the types you will see most often. We'll present these types of constraints and objectives in the following solved problems.

Modeled Optimization Problems

Example 1: Green Infrastructure
A company has budgeted $45,000 to capture stormwater through the use of a green roof and cisterns. A modular green roof system costs $50 per unit, which covers 2.6 square feet of area and stores 9 gallons. The Cistern costs $680 per unit and stores 500 gallons. While the cistern is 48 inches in diameter, with room for access and plumbing, the footprint of each cistern is actually 25 square feet. The office building has room for no more than 2000 square feet on the roof and 295 square feet on land to house the cisterns. The green roof manufacturer has a minimum order quantity of 760 units, and a maximum order quantity of 3000 units. Assuming the building is structurally able to handle the weight of the green roof system, how many cisterns and how much green roof should you buy in order to maximize

storage volume?

Decision Variables:

C: Number of Cisterns

G: Number of green roof modules

Y: Binary Indicator Variable = 1 if green roof is being used, 0 if not

Objective Function:

Maximize Volume = 9 G + 500 C

Subject to:

Constraints:

Green Roof Area: 2.6 G ≤ 2000

Cistern Area: 25 C ≤ 295

Budget: 50 G + 680 C ≤ 45,000

Minimum Order: G ≥ 760 Y or G -760 Y ≥ 0

Maximum Order: G ≤ 3000 Y or G – 3000 Y ≤ 0

Binary: Y is 0 or 1. Notice that if Y is 0, then both the minimum and maximum quantity order constraints will be satisfied as G will also be 0. If Y is 1 (indicating that green roof modules will be used), then G will be forced to be between 760 and 3000).

Example 2: Renewable Diesel Production from Biosolids
At a biosolids energy recovery facility using pyrolysis, the refining process requires the production of at least two gallons of diesel for each gallon of aviation fuel. To meet the anticipated demands of the facility's customers, at least 3000 gallons of aviation fuel a day will need to be produced. The demand for diesel, on the other hand, is not more than 6400 gallons a day.

If diesel is selling for $3.90 per gallon and aviation fuel sells for $3.50/gal, how much of each should be produced in order to maximize revenue?

Decision Variables:

D: Amount of diesel produced

A: Amount of aviation fuel produced

Objective Function:

Max 3.9 D + 3.5 A

Subject to:

Constraints:

Demands

Aviation fuel demand: $A \geq 3000$

Diesel demand: $D \leq 6400$

Process limitations: $D > 2A$ or $2A - D \leq 0$

Example 3: Blending Water Sources for Reuse
A reclaimed water provider makes two types of water, ultrapure from a reverse osmosis (RO) process, and high quality treated effluent from a nitrification plant. The water quality from each is shown in the table below:

Water Sources	TDS (mg/L)	Silica (mg/L)	Total Nitrogen (mg/L)
Ultrapure RO	5	0.1	1.0
Nitrification Plant	2000	35	7.5

The provider serves two customers. The first is an irrigation customer that needs a minimum of 570 m3 per day (150,000 gallons per day) and has minimum quality requirements for Total Dissolved Solids (TDS) and Total Nitrogen. The second is an industrial facility that requires 380 m³ per day (100,000 gallons per day) for boiler makeup. The boiler water has minimum quality constraints for TDS and Silica. The customer water quality demands are shown below:

Water Reuse Customers	TDS (mg/L)	Silica (mg/L)	Total Nitrogen (mg/L)
Irrigation	450	No defined limit	5
Boiler at industrial facility	60	1.5	No defined limit

How much water should be treated by each of the systems (the ultrapure RO system and the Nitrification plant) in order to blend the two sources to make product acceptable to the two customers? There's no limit on how much water the provider can produce, as any excess production beyond the demands of the irrigation and boiler customers is used for aquifer recharge to combat salt water intrusion. Because the operations of the RO system are substantially more expensive than the nitrification plant, the goal is to minimize the amount of water treated by the RO system, while meeting the demands (both quantity and quality) of the two customers.

Decision Variables:

UI: Ultrapure volume used for irrigation

NI: Nitrification plant volume used for irrigation

UB: Ultrapure volume used for boilers

NB: Nitrification plant volume used for boilers

Objective Function:

Minimize UI+UB

Subject to:

Constraints:

TDS at Irrigation: $5\ UI + 2000\ NI \leq 450\ (UI+NI)$

Nitrogen at Irrigation: $1\ UI + 7.5\ NI \leq 5\ (UI+NI)$

TDS at Boiler: $5\ UB + 2000\ NB \leq 60\ (UB+NB)$

Silica at Boiler: $0.1\ UB + 35\ NB \leq 1.5\ (UB+NB)$

Boiler Demand: $UB + NB \geq 380$

Irrigation Demand: $UI + NI \geq 570$

Example 4: Customer Service Staffing
During the summer months, the City of Mountain View staffs customer service centers seven days a week to meet the demands of the vacationing crowds. Regulations require that city employees work five days a week and be given two consecutive days off. The city manager's policy to ensure an acceptable level of service mandates that Mountain View provide at least one representative per 80 average calls per day. The average daily calls from last summer are as follows: Sunday – 580, Monday – 420, Tuesday – 350, Wednesday – 250, Thursday – 440, Friday – 510 and Saturday – 680. Given a tight budget constraint, the city would like to determine a schedule that will employ as few representatives as possible. How many should the city hire and what schedule should they work?

Decision Variables:

CSR1: Customer Service Rep who starts their 5 day schedule on Sunday
CSR2: Customer Service Rep who starts their 5 day schedule on Monday
CSR3: Customer Service Rep who starts their 5 day schedule on Tuesday
CSR4: Customer Service Rep who starts their 5 day schedule on Wednesday
CSR5: Customer Service Rep who starts their 5 day schedule on Thursday
CSR6: Customer Service Rep who starts their 5 day schedule on Friday
CSR7: Customer Service Rep who starts their 5 day schedule on Saturday

Objective Function:

Min CSR1 + CSR2 + CSR3 + CSR4 + CSR5 +CSR6 + CSR7

Subject to:

Constraints:

Sunday: CSR1 + CSR4 + CSR5 +CSR6 + CSR7 ≥ 8 (580/80 rounded up to whole person)

Monday: CSR1 + CSR2 + CSR5 +CSR6 + CSR7 ≥ 6 (420/80 rounded up to whole person)

Tuesday: CSR1 + CSR2 + CSR3 + CSR6 + CSR7 ≥ 5 (350/80 rounded up to whole person)

Wednesday: CSR1 + CSR2 + CSR3 + CSR4 + CSR7 ≥ 4 (250/80 rounded up to whole person)

Thursday: CSR1 + CSR2 + CSR3 + CSR4 + CSR5 ≥ 6 (440/80 rounded up to whole person)

Friday: CSR2 + CSR3 + CSR4 + CSR5 +CSR6 ≥ 7 (510/80 rounded up to whole person)

Saturday: CSR3 + CSR4 + CSR5 +CSR6 + CSR7 ≥ 9 (680/80 rounded up to whole person)

All variables integers, as you can't have fractional people.

Example 5: Nutrient Recovery From Biosolids
Nutrient management planning ensures that the appropriate quantity and quality of biosolids are land applied to farmland. The biosolids application is specifically calculated to match the nutrient uptake requirements of the particular crop. Nutrient management technicians work with the farm community to assure proper land application and nutrient control.

The Dos Rios Water Resource Recovery Facility produces two types of fertilizer products from its biosolids, a Class A product (containing no detectible levels of pathogens) and a Class B product (highly treated but still containing detectible levels of pathogens). There are buffer requirements, public access, and crop harvesting restrictions. These two products can be used by different customers. It costs \$1.75 to produce a bag of Class A product and \$1000 to produce a load of Class B, which is equivalent in volume to 1000 bags of Class A. In order to comply with permits, the WRRF has a policy that at least 30%, but not more than 60%, of its output volume must be Class B.

The plant wants to meet but not exceed demand for each product. The marketing manager estimates that the maximum demand for Class B is 5 loads. Market research indicates that each \$1000 spent on advertising of the Class B product will increase sales by one load. Maximum demand for Class A is estimated to be 4000 bags, plus an additional 5 bags for every \$1 spent to promote the Class A product. The company has \$160,000 to spend on producing and advertising their fertilizers. Class A sells for \$3.00 per bag; Class B sells for \$2500 per load. The organization wants to know how many units of each to produce and how much advertising to spend on each in order to maximize profit.

Decision Variables:

A: Number of bags of Class A produced

B: Number of loads of Class B produced

MA: Marketing costs for advertising Class A product

MB: Marketing costs for advertising Class B product

Objective Function:

Max Revenues –Total Costs

Subject to

Constraints:

Revenues $= 3A + 2500B$

Total Costs $=$ Marketing $+$ Production

Marketing = MB + MA

Production = 1.75A + 1000B

A = 4000 + 5MA (Production of Class A)

B = 5 + 0.001MB (Production of Class B)

B ≥ 0.3 (1000B + A) (Lower limit of at least 30% Class B)

B ≤ 0.6 (1000B +A) (Upper Limit of no more than 60% Class B)

Total Costs ≤ 160,000 (Budget Limit)

Example 6: Job Assignments
Appletree Water Plant has four workers working 12-hour shifts. To complete the day's work they must perform regular mechanical repair, laboratory sampling and testing, electrical maintenance, and reporting functions, as well as be prepared for other tasks that arise each day. The time it takes each worker to do each job is shown below. Each worker is assigned one primary job and serves as a floater to assist on general tasks that arise during the normal day. Determine the assignments which complete the tasks in the quickest total time, therefore leaving more time for the entire staff to address other work.

Hours/job by Employee	Mechanical	Laboratory	Electrical	Reporting
Jose	6	5	2	1
Damon	9	8	7	3
Shelly	8	5	9	4
Tyler	7	7	8	3

Decision Variables:

JM: 1 if Jose does Mechanical Job, 0 if not
DM: 1 if Damon does Mechanical Job, 0 if not
SM: 1 if Shelly does Mechanical Job, 0 if not
TM: 1 if Tyler does Mechanical Job, 0 if not

Objective Function:

Minimize $J + D + S + T$

Subject to:

Constraints:

Time for each job per worker

$J = 6\,JM + 5JL + 2JE + 1\,JR$
$D = 9\,DM + 8DL + 7\,DE + 3\,DR$
$S = 8\,SM + 5\,SL + 9\,SE + 4\,SR$
$T = 7\,TM + 7\,TL + 8\,TE + 3\,TR$

Each worker gets one job

$JM + JL + JE + JR = 1$ (Jose's assignment)
$DM + DL + DE + DR = 1$ (Damon's assignment)
$SM + SL + SE + SR = 1$ (Shelly's assignment)
$TM + TL + TE + TR = 1$ (Tyler's assignment)

Each job gets one worker

$JM + DM + SM + TM = 1$ (Mechanical)
$JL + DL + SL + TL = 1$ (Laboratory)
$JE + DE + SE + TE = 1$ (Electrical)
$JR + DR + SR + TR = 1$ (Reporting)
All variables Binary (0 or 1)

Example 7: Biogas Production

A regional group has consolidated the management and production of biogas. The group has three sources of supply and three wholesale customers. Due to differences in gas quality and transportation costs, the profit to the group is related to which source is bought by which customer. Formulate a transportation problem to maximize the group's profit given the following cost, supply, and demand data. If there is not enough gas for a customer, the customer can purchase the shortfall from another supplier.

Capacities:

Landfill Storage Facility = 50 (thousand cubic feet)

WRRF Storage Facility = 100 (thousand cubic feet)

Agricultural Storage Facility = 50 (thousand cubic feet)

Maximum Demand:

Power Plant = 80 (thousand cubic feet)

Bus Fleet = 90 (thousand cubic feet)

Drying Kiln = 100 (thousand cubic feet)

Profit per unit	Power Plant	Bus Fleet	Drying Kiln
Landfill	69	60	75
WRRF	79	73	68
Agricultural	85	76	70

NW CORNER METHOD

One way to find a feasible solution without using operations research techniques is what's called the "Northwest Corner Method." Using this method you assign quantities starting at the top left of the matrix and make sure you meet the demand and supply constraints of the top leftmost cell (the Northwest Corner). Then you methodically work your way down and right until you have covered all supplies and demands. This approach provides a workable solution, and a profit of $14,550 in this example, as shown in the table below.

Profit per unit	Power Plant	Bus Fleet	Drying Kiln	Max. Capacity
Landfill	50 $69	0 $60	0 $75	50
WRRF	30 $79	70 $73	0 $68	100
Agricultural	0 $85	20 $76	30 $70	50
Maximum Demand	80	90	100	

While we can easily find a solution without LP, using the tool can definitely make us some money. The LP model is shown below.

Decision Variables:

LP: Volume from Landfill to Power Plant
LB: Volume from Landfill to Bus Fleet
LD: Volume from Landfill to Drying Kiln
WP: Volume from WRRF to Power Plant
WB: Volume from WRRF to Bus Fleet
WD: Volume from WRRF to Drying Kiln
AP: Volume from Agricultural to Power Plant
AB: Volume from Agricultural to Bus Fleet
AD: Volume from Agricultural to Drying Kiln

Objective Function:

Maximize Profit

Subject to:

Constraints:

Profit = Landfill + WRRF + Agricultural
Landfill = 69 LP + 60 LB + 75 LD
WRRF = 79 WP + 73 WB + 68 WD
Agricultural = 85 AP + 76 AB + 70 AD

Supply:

Landfill = LP + LB + LD $\leq$ 50

WRRF = WP + WB + WD $\leq$ 100

Agricultural = AP + AB + AD $\leq$ 50

Demand:

Power Plant = LP + WP + AP $\leq$ 80

Bus Fleet = LB + WB + AB $\leq$ 90

Drying Kiln = LD + WD + AD $\leq$ 100

The optimized solution from the Linear Program is shown below. Profit is $15,480, an increase of nearly $1000.

Profit per unit	Power Plant	Bus Fleet	Drying Kiln	Max. Capacity
Landfill	0 $69	0 $60	50 $75	50
WRRF	30 $79	70 $73	0 $68	100
Agricultural	50 $85	0 $76	0 $70	50
Maximum Demand	80	90	100	

Example 8: O&M Team Logistics

NOVA Water Company has 4 trucks for their O&M teams. Each truck is taken home by the team leads so they can proceed directly to the jobsite in the morning. Jobsites for the day are at Shady Pines, Oak Hill, Elm View, and Willowbrook. The costs for getting each truck to each site are below. What sites should each truck go to in order to minimize the utility's cost?

Location	Truck 1 Cost	Truck 2 Cost	Truck 3 Cost	Truck 4 Cost
Shady Pines	$160	$20	$130	$30
Oak Hill	$60	$140	$30	$70
Elm View	$70	$90	$120	$110
Willowbrook	$40	$110	$60	$50

Like the Northwest Corner method used in the previous gas transportation problem, this type of assignment problem can also be set up by starting with the lowest cost and making assignments on the lowest cost remaining until all trucks are assigned to locations. The lowest cost is $20 to get Truck 2 to Shady Pines, so we start there. Then we move to $30 for Truck 3 to go to Oak Hill, $40 to get Truck 1 to Willowbrook, leaving Truck 4 to service Elm View at $110 for a grand total of $200 in the baseline solution below.

Location	Truck 1 Cost	Truck 2 Cost	Truck 3 Cost	Truck 4 Cost
Shady Pines	$160	**$20**	$130	$30
Oak Hill	$60	$140	**$30**	$70
Elm View	$70	$90	$120	**$110**
Willowbrook	**$40**	$110	$60	$50

Decision Variables:

S1: 1 if Truck 1 goes to Shady Pines, 0 otherwise
O1: 1 if Truck 1 goes to Oak Hill, 0 otherwise
E1: 1 if Truck 1 goes to Elm View, 0 otherwise
W1: 1 if Truck 1 goes to Willowbrook, 0 otherwise
S2: 1 if Truck 2 goes to Shady Pines, 0 otherwise
O2: 1 if Truck 2 goes to Oak Hill, 0 otherwise
E2: 1 if Truck 2 goes to Elm View, 0 otherwise
W2: 1 if Truck 2 goes to Willowbrook, 0 otherwise
S3: 1 if Truck 3 goes to Shady Pines, 0 otherwise
O3: 1 if Truck 3 goes to Oak Hill, 0 otherwise

23

E3: 1 if Truck 3 goes to Elm View, 0 otherwise
W3: 1 if Truck 3 goes to Willowbrook, 0 otherwise
S4: 1 if Truck 4 goes to Shady Pines, 0 otherwise
O4: 1 if Truck 4 goes to Oak Hill, 0 otherwise
E4: 1 if Truck 4 goes to Elm View, 0 otherwise
W4: 1 if Truck 4 goes to Willowbrook, 0 otherwise

Objective Function:

Minimize S + O + E + W

Subject to:

Constraints:

Summary cost variable:

S = 160 S1 +20 S2 + 130 S3 + 30S4
O= 60 O1 + 140 O2 + 30 O3 + 70 O4
E = 70 E1 + 90 E2 + 120 E3 + 110 E4
W = 40 W1 +110 W2 + 60 W3 + 50 W4

Each site gets one truck:

S1 + S2 + S3 + S4 = 1
O1 + O2 +O3 + O4 = 1
E1 + E2 + E3 + E4 = 1
W1 + W2 + W3 + W4 = 1

Each truck goes to one site:

S1 + O1 + E1 + W1 = 1
S2 + O2 + E2 + W2 = 1
S3 + O3 + E3 + W3 = 1
S4 + O4 + E4 + W4 = 1
All variables Binary.

The optimized solution is shown below, yielding a cost of $170, representing a savings of 15 percent.

Location	Truck 1 Cost	Truck 2 Cost	Truck 3 Cost	Truck 4 Cost
Shady Pines	$160	$20	$130	$30
Oak Hill	$60	$140	$30	$70
Elm View	$70	$90	$120	$110
Willowbrook	$40	$110	$60	$50

Example 9: Transmission and Distribution

A new reservoir is being developed in Appalachia to transmit and distribute (T&D) water to Bristow Falls. The T&D network has pump stations and pipe networks with the capacities shown by the network diagram in Figure 2 below (in 000s m³/day). Develop the LP problem to find the schedule that gets the most water to Bristow Falls. This is what is typically referred to as a "Max Flow" problem.

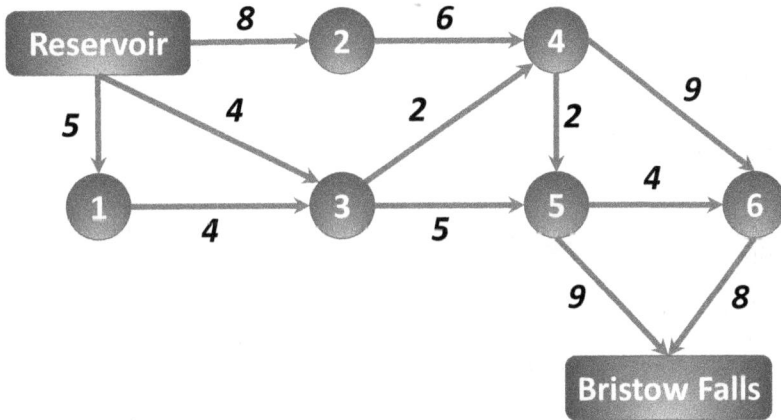

Figure 2 Reservoir T&D network diagram.

Decision Variables:

RP1: Amount from Reservoir to Pump Station 1
RP2: Amount from Reservoir to Pump Station 2

25

RP3: Amount from Reservoir to Pump Station 3

P1P3: Amount pumped from Pump Station 1 to Pump Station 3

P2P4: Amount pumped from Pump Station 2 to Pump Station 4

P3P4: Amount pumped from Pump Station 3 to Pump Station 4

P3P5: Amount pumped from Pump Station 3 to Pump Station 5

P4P5: Amount pumped from Pump Station 4 to Pump Station 5

P4P6: Amount pumped from Pump Station 4 to Pump Station 6

P5P6: Amount pumped from Pump Station 5 to Pump Station 6

P5B: Amount from Pump Station 5 to Bristow Falls

P6B: Amount from Pump Station 6 to Bristow Falls

Objective Function:

Max P5B + P6B (all of the inputs to Bristow Falls) or

Max RP1 + RP2 + RP3 (all of the outputs from the Reservoir)

Subject to:

Constraints:

Pipe capacities

RP3 $\leq$ 4

P2P4 $\leq$ 6

P3P4 $\leq$ 2

P1P3 $\leq$ 4

P3P5 $\leq$ 5

P4P5 $\leq$ 2

P4P6 $\leq$ 9

P5P6 $\leq$ 4

P5B $\leq$ 9

P6B $\leq$ 8

Balance at intermediate nodes (total in = total out)

At Pump Station 1: RP1 = P1P3

At Pump Station 2. RP2 = P2P4

At Pump Station 3: P1P3 + RP3 = P3P4 + P3P5

At Pump Station 4: P2P4 + P3P4 = P4P5 + P4P6
At Pump Station 5: P3P5 + P4P5 = P5B + P5P6
At Pump Station 6: P4P6 + P5P6 = P6B

Example 10: Truck Routing

If we used the same network diagram, but view it as a set of streets instead of pipes, we can demonstrate the shortest path approach (see Figure 3 below). Consider that a truck needs to leave the operations depot to perform a repair job at a site. Each road takes a certain amount of time to travel. For example, it takes six minutes to travel from intersection 2 to intersection 4. Each node, which was a pumping station in the previous example, is now an intersection. What is the shortest route to get from the operations depot to the jobsite? Note that this is the type of analysis that FedEx, UPS, and other logistical companies perform every day.

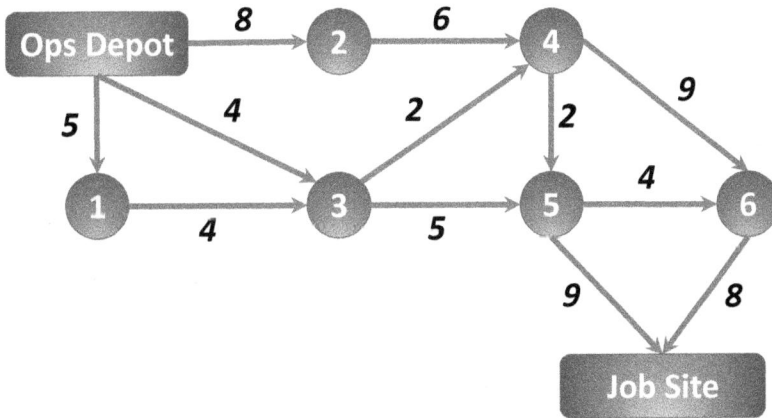

Figure 3 Truck route network diagram.

Decision Variables:

T1: Time to travel to Intersection 1
T2: Time to travel to Intersection 2
T3: Time to travel to Intersection 3
T4: Time to travel to Intersection 4

T5: Time to travel to Intersection 5

T6: Time to travel to Intersection 6

TJobsite: Time to travel to Jobsite

TOpsDepot: Time to travel to Operations Depot (or 0)

Objective Function:

Min TJobsite

Subject to:

Constraints:

Node 1

TOpsDepot+ 5 ≤ T1

Node 2

TOpsDepot + 8 ≤ T2

Node 3:

TOpsDepot + 4 ≤ T3

T1+4 ≤ T3

Node 4:

T2+6 ≤ T4

T3+2 ≤ T4

Node 5:

T4+2 ≤ T5

T3+5 ≤ T5

Node 6:

T4+9 ≤ T6

T5+4 ≤ T6

Jobsite:

T5+9 ≤ TJobsite

T6+8 ≤ TJobsite

Operations Depot Starting Point

TOpsDepot = 0

Example 11: Reservoir Operations
A water utility has a reservoir that is primarily used for flood control, but the O&M costs are offset by wholesale raw water sales to customers. The location is in an area that receives very little rain in the summer when temperatures are hot and increase evaporation from the surface of the lake. The rainy season is in the winter, when evaporation is low. The demand for water increases in the summer, which allows the owner to sell water at $3.00 per m3 from April through October, but only for $1.50 per m³ from November through March. On January 1 of each year, there must be a minimum of 2 million m³ in the reservoir, and every other month the volume can drop to 90 percent of this value (or 1.8 million m³). Assume that the weather history is consistent enough that you are fairly certain of the rainfall and evaporation forecast, as shown in Figure 4 below. What program of selling water (how much water should you sell in each month) will maximize revenues, while meeting the minimum volume in the lake?

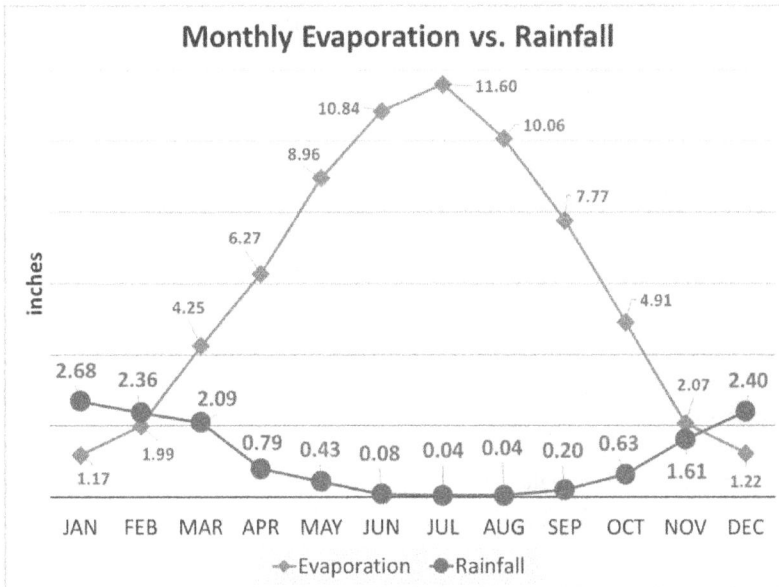

Figure 4 Rainfall and evaporation forecast.

After converting from inches to mm, and considering that 1 mm/month of rain corresponds to 10 m³/month per hectare of the watershed, the following table shows the evaporation and inflows in the watershed.

	Evaporation (mm/mo.)	Evaporation from 500 hectare reservoir (m³/month)	Rainfall (mm/month)	Inflows from Rainfall in 20,000 hectare watershed assuming runoff factor of 0.5 (m³/month)
Jan	30	14,859	68	680,720
Feb	51	25,273	60	599,440
Mar	108	53,975	53	530,860
Apr	159	79,629	20	200,660
May	228	113,792	11	109,220
Jun	275	137,668	2	20,320
Jul	295	147,320	1	10,160
Aug	256	127,762	1	10,160
Sep	197	98,679	5	50,800
Oct	125	62,357	16	160,020
Nov	53	26,289	41	408,940
Dec	31	15,494	61	609,600

Decision Variables:

Sales1: Volume of water sold during Month 1
Reservoir1: Volume of water in reservoir at end of Month 1
Sales2: Volume of water sold during Month 2
Reservoir2: Volume of water in reservoir at end of Month 2
And so on…

Objective Function:

Max Revenue = $1.50 * (Sales1 + Sales2 + Sales3 + Sales11 + Sales12) + $3.00 * (Sales4 + Sales5 + Sales6 + Sales7 + Sales8 + Sales9 + Sales10)

Subject to:

Constraints:

Water Balance (volume in the lake at the beginning of the month minus the evaporation during the month plus the rainfall during the month minus the sales during the month equals the volume in the lake at the end of the month)

January (1): Reservoir0 -14,859 + 680,720 – Sales1 = Reservoir1
February (2): Reservoir1 - 25,273 + 599,440 – Sales2 = Reservoir2
March (3): Reservoir2 - 53,975 + 530,860 – Sales3 = Reservoir3
April (4): Reservoir3 -79,629 + 200,660 – Sales4 = Reservoir4
May (5): Reservoir4 -113,792 + 109,220 – Sales5 = Reservoir5
June (6): Reservoir5 -137,668 + 20,320 – Sales6 = Reservoir6
July (7): Reservoir6 -147,320 + 10,160 – Sales7 = Reservoir7
August (8): Reservoir7 - 127,762 + 10,160 – Sales8 = Reservoir8
September (9): Reservoir8 -98,679 + 50,800 – Sales9 = Reservoir9
October (10): Reservoir9 -62,357 + 160,020 – Sales10 = Reservoir10
November (11): Reservoir10 -26,289 + 408,940 – Sales11 = Reservoir11
December (12): Reservoir11 -15,494 + 609,600 – Sales12 = Reservoir12

Minimum Volume in Reservoir

Reservoir0 = 2,000,000
Reservoir1 ≥ 1,800,000
Reservoir2 ≥ 1,800,000
Reservoir3 ≥ 1,800,000
Reservoir4 ≥ 1,800,000
Reservoir5 ≥ 1,800,000
Reservoir6 ≥ 1,800,000
Reservoir7 ≥ 1,800,000
Reservoir8 ≥ 1,800,000
Reservoir9 ≥ 1,800,000
Reservoir10 ≥ 1,800,000
Reservoir11 ≥ 1,800,000
Reservoir12 ≥ 2,000,000

Recommended References

AMPL Optimization website: http://ampl.com/
General Algebraic Modeling System (GAMS) website:
http://www.gams.com/
OptaPlanner website: http://www.optaplanner.org/
Optimatics website: http://optimatics.com/

PART II: INSIDE THE BLACK BOX

3 INTRODUCTION TO OPTIMIZATION TECHNIQUES

Part I showed how a seemingly complex problem can be structured rather straightforwardly into an optimization problem. Then the "black box" takes over and runs the calculations. While this text will not go into the intricate mathematics of those calculations, we will take a quick look at the mathematical constructs that go into the solver, analytical engine, or whatever other technical term could be used for the "black box."

Most people have been performing mathematical optimization more often than they realize. Some techniques have been taught in middle school and high school. We'll graphically demonstrate the simple application of algebraic and differential calculus approaches here, simply to prove that you have probably seen some type of optimization before. The following two chapters will discuss two approaches commonly used in operations research applications: Linear Programming and Evolutionary Algorithms.

Algebraic Optimization

Suppose a contractor needs to build a concrete tank. Based on the location and use, the designer has specified that the concrete aggregate blend must have between 42 and 45 percent sand. The contractor has two suppliers of aggregate, Albert's Supply and Bart's Quarry, that they could blend. Albert has prepackaged aggregate that has 35% sand and 65% gravel and sells for $5.00 per cubic meter. Bart has a similar package that has 45% sand and 55% gravel and sells for $6.00 per cubic meter. The obvious solution is to use as much of the cheaper concrete from Albert so that the blend barely makes it into the 42-45% range. The algebraic setup is as follows:

Let x = the percentage of concrete in the blend from Albert's Supply

Let 1- x = the percentage of concrete in the blend from Bart's Quarry

Then solve for x using the sand balance:

x (0.35) + (1- x)(0.45) = 0.42
0.35 x + 0.45 – 0.45 x = 0.42
0.35 x – 0.45 x = 0.42 – 0.45
- 0.1 x = -0.03
x = 0.3

As shown in the graph in Figure 5, the blended aggregate will contain:

30% of the \$5.00/m³ material from Albert

70% of the \$6.00/m³ material from Bart

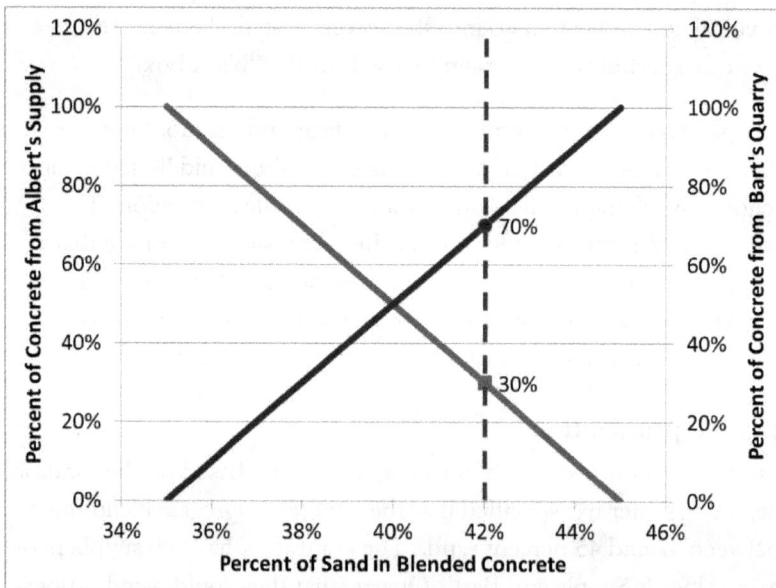

Figure 5 Aggregate blending.

The lowest cost per cubic meter of blended aggregate is:

0.30 x $5.00 + 0.70 x $6.00 = 1.50 + 4.20

= $5.70 per cubic meter

Differential Calculus

Another type of optimization is demonstrated by a classical physics problem: maximum height of a projectile. For example, let's assume you launch a model rocket straight up on a windless day (no movement other than up and down). The launch velocity is 29.4 meters per second. What will be the maximum height attained by the rocket?

The standard equation for the height of a projectile is:

$$h = v_0 t - \tfrac{1}{2}gt^2$$

where h = height at time t, v_0 = initial velocity, g = gravitational acceleration (~9.8 m/s²), and t = time in seconds.

So in this case, the h= $29.4t - \tfrac{1}{2}(9.8)t^2$. In order to find the time at which the height is a maximum, we set the first derivative to 0. The first derivative of h is

$$h' = 29.4 - 9.8t$$

Setting h' = 0 yields

$$0 = 29.4 - 9.8t$$

$$9.8t = 29.4$$

$$t = 3$$

Therefore, the maximum height occurs 3 seconds after launch. Plugging into the original equation yields

$h = 29.4 \ (3) - \frac{1}{2}(9.8)(32)$

$h = 88.2 - 44.1$

$h = 44.1$

Therefore , maximum height is 44.1 meters, as shown in the graph in Figure 6 below.

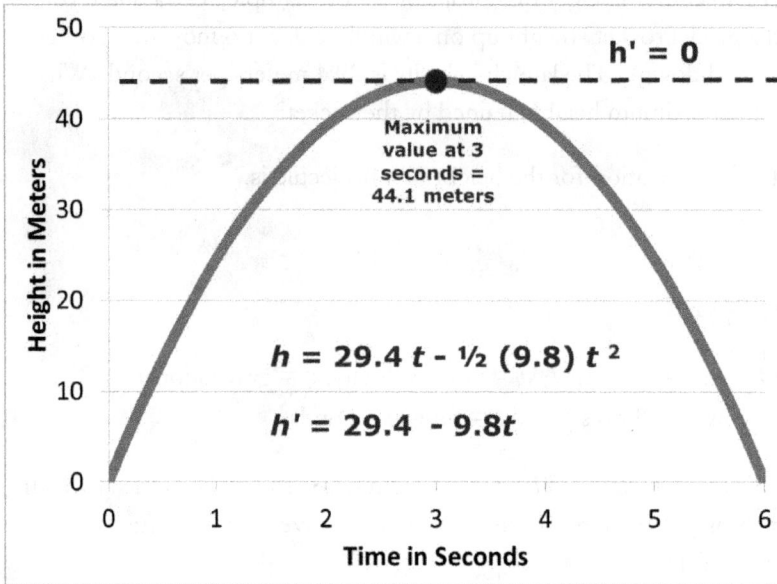

Figure 6 Maximum height of a projectile.

4 LINEAR PROGRAMMING

A Linear Programming model seeks to maximize or minimize a linear function, subject to a set of linear constraints. Recall from Chapter 2 that the linear model consists of the following components:

- A set of decision variables.
- An objective function.
- A set of constraints.

Consider the following problem:

Nitrogen and phosphorus are critical nutrients for agriculture, and land application of Class A biosolids can be an effective and natural way to recycle nutrients into the soil. At the Xavier biosolids processing facility, each load of biosolids contains 2 tons of nitrogen and 1 ton of phosphorus. At the Yellow Springs facility, each load contains 2 tons of nitrogen and 2 tons of phosphorus. A farming collective wants to blend the biosolids for land application as natural fertilizer on their fields. The maximum amount of nitrogen the fields can handle is 9 tons and the maximum phosphorus load is 7 tons. The net revenue per load for the biosolids processor Xavier is $300 and $400 at Yellow Springs. In order to maximize the revenues to the biosolids processor, how many loads from each facility should be provided? For simplicity, assume that any additional biosolids production is being recycled in a different manner.

In plain English, the objective is to maximize the net revenue to the water resource recovery facility. The decision variables are how much of the

37

biosolids production at each plant (Xavier and Yellow Springs) should be allocated to land application. The limiting factors are that the agricultural fields can receive a maximum of 9 tons of nitrogen and 7 tons of phosphorus.

Converting to Mathematical Model

The decision variables are the number of tons processed at each facility. Let's set the following variables:

- X = number of loads produced by Xavier facility
- Y = number of loads produced by Yellow Springs facility

The objective function is to maximize net revenue. Total revenues are calculated by multiplying the unit price times the number of loads sold:

- Maximize Total Net Revenues = $300X + 400Y$

There are two constraints: phosphorus demand and nitrogen demand. The amount of each of the nutrients removed varies by facility, with the concentrations specified in the problem.

- Phosphorus demand constraint = 1 ton of phosphorus produced in a load at Xavier facility. Similarly, at Yellow Springs, each load contains 2 tons of phosphorus. The total of these must be less than or equal to 7 tons in order to meet the nutrient demands without providing too much which could lead to pollutant runoff. Thus the constraint can be written as:

$$X + 2Y \leq 7$$

- Following the same approach for the other nutrient makes the equation for the nitrogen constraint:

$$2X + 2Y \leq 9$$

Translate to format for analytical engine:

Maximize Total Net Revenues = 300X + 400Y

s.t. (subject to the constraints)

Phosphorus: X + 2Y ≤ 7

Nitrogen: 2X + 2Y ≤ 9

When a linear program only contains two variables, we can solve it graphically, which will illustrate the techniques that can be expanded to larger problems by using an analytical engine. Let's solve this problem graphically using Linear Programming.

First, let's graph the constraints. Figure 7 below shows the nitrogen constraint (2X + 2Y ≤ 9). The shaded area shows where this constraint is met.

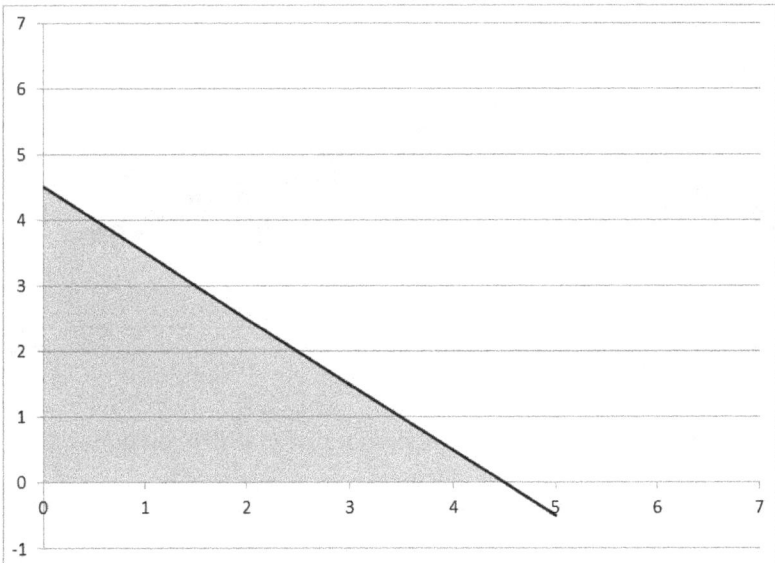

Figure 7 Nitrogen constraint.

Now we add the phosphorus constraint (X + 2Y ≤ 7). This is shown in
Figure 8 below by the line going diagonally from 3.5 on the vertical axis to 7
on the horizontal axis. There are some areas that meet the phosphorus
constraint but don't meet the nitrogen constraint. Another area shows
potential solutions that meet the nitrogen constraint, but not the
phosphorus constraint. These infeasible areas are highlighted in black. The
remaining grey area is called the **feasible region**, the set of potential
solutions where all constraints are met. Please note that this also assumes
that the values of X and Y must also be greater than or equal to zero.
While this non-negativity constraint is not a requirement of an optimization
problem, they often exist.

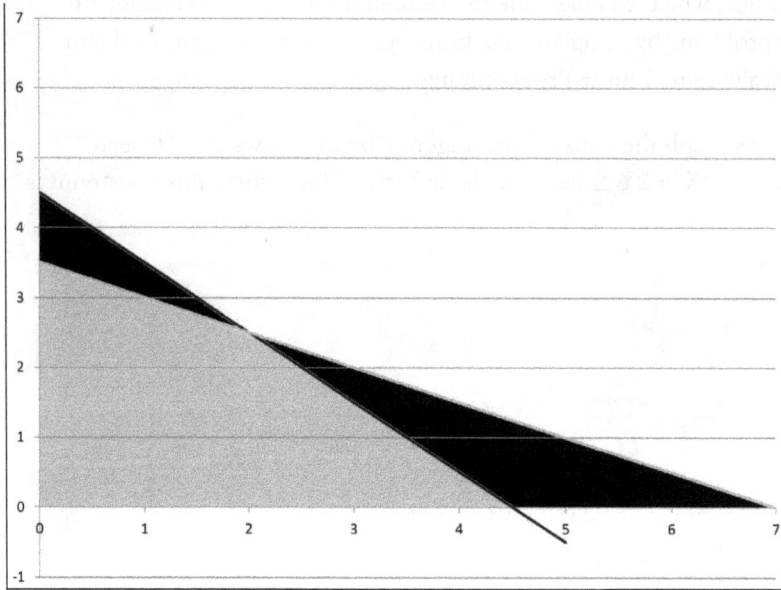

Figure 8 Addition of Phosphorus constraint to Nitrogen constraint.

Now that we have our potential solutions that meet our constraints (the feasible region), we have to put our objective function on the graph (see below in Figure 9). The dotted line represents a line that follows the equation Total Net Revenues = 300X + 400 Y. In this case, the total net revenues are $400. This is a starting point.

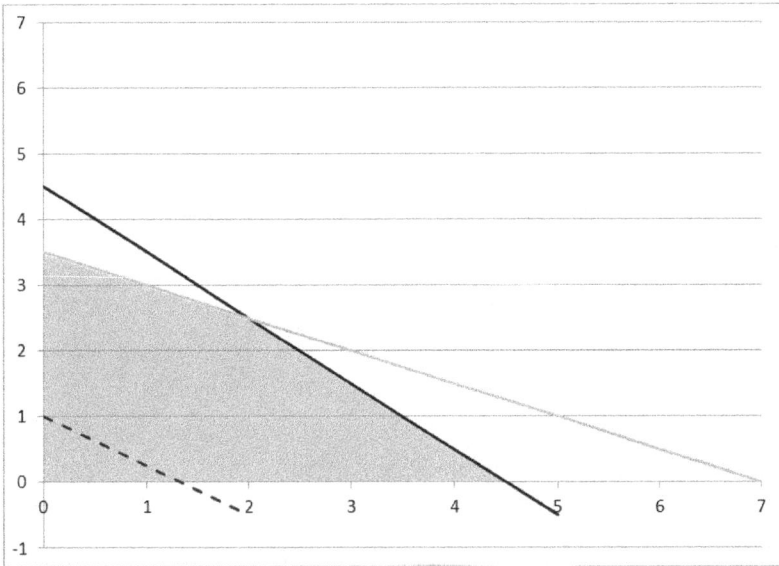

Figure 9 Objective function.

Now that we have a potential solution that will get us $400, let's see how we can make more money. Let's increase the Total Net Revenues to $1200 and see if we can still meet our constraints. The graph in Figure 10 below shows that our objective function line is still in the feasible region, so we can pick a solution that makes us $1200. But there's room for more.

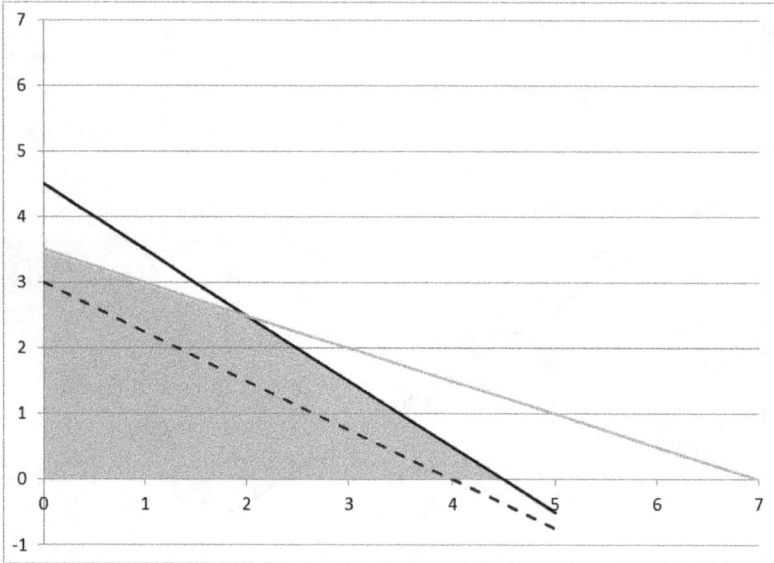

Figure 10 Increasing Total Net Revenues to $1200.

If we keep sliding the objective function line upward, we will eventually reach the edge of the feasible region. As shown in the graph in Figure 11 below, we've touched the extreme edge of the feasible region at point (2, 2.5). That is, with 2 loads from Xavier and 2.5 loads from Yellow Springs, we still meet all constraints. The Total Net Revenues from this combination are $1600 ($300 x 2 loads from Xavier + $400 x 2.5 loads from Yellow Springs). This is our Maximum Total Net Revenues. As you might guess, the answer to optimization problems are found at the edges of the feasible region. These are known as **extreme points**.

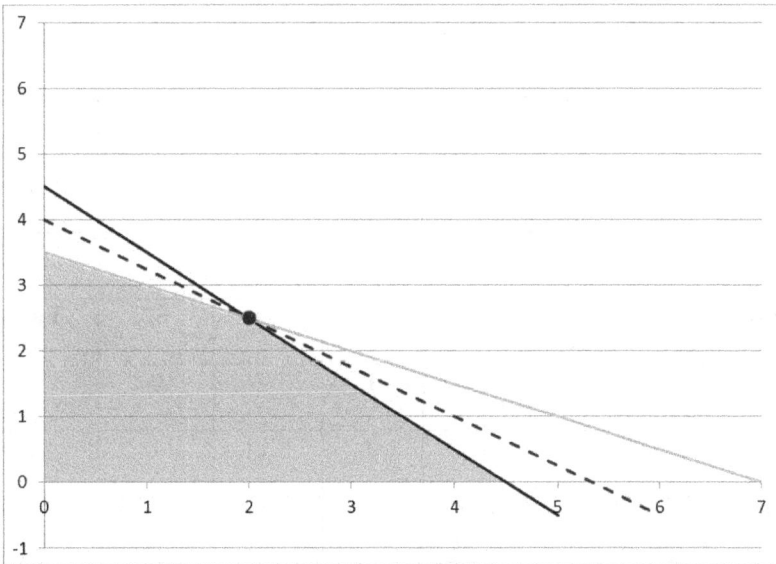

Figure 11 Solution Found at an Extreme Point.

The tools that take this same approach on more than just two variables use various solution algorithms. Probably the most well-known algorithm is the Simplex Method, a matrix-based approach to evaluating the extreme points to get to the optimal solution. This example problem, when solved in the Simplex Method, would yield the following sets of result matrices (called tableaus) on the way to a solution. (We're not going to go into the Simplex Method here, but just use these tableaus to show how the mathematical engine would come up with the same answer as the graphical solution.)

Initial Tableau	X1	X2	S1	S2	Value
S1	1	2	1	0	7
S2	2	2	0	1	9
Z (objective)	-300	-400	0	0	0

Iteration 1	X1	X2	S1	S2	Value
S1	0.5	1	0.5	0	3.5
S2	1	0	-1	1	2
Z (objective)	-100	0	200	0	1400

Final Tableau	X1	X2	S1	S2	Value
Xavier	0	1	1	-0.5	2.5
Yellow Springs	1	0	-1	1	2
Total Net Revenues	0	0	100	100	1600

So we've solved our problem and can maximize our net revenues if we take 2 loads from Xavier and 2.5 loads from Yellow Springs. What if we can't send half loads? We can modify the linear program to limit our solutions to whole numbers. This approach is called integer programming. The feasible region under integer programming is all the whole number points in the original feasible region, as shown below in Figure 12.

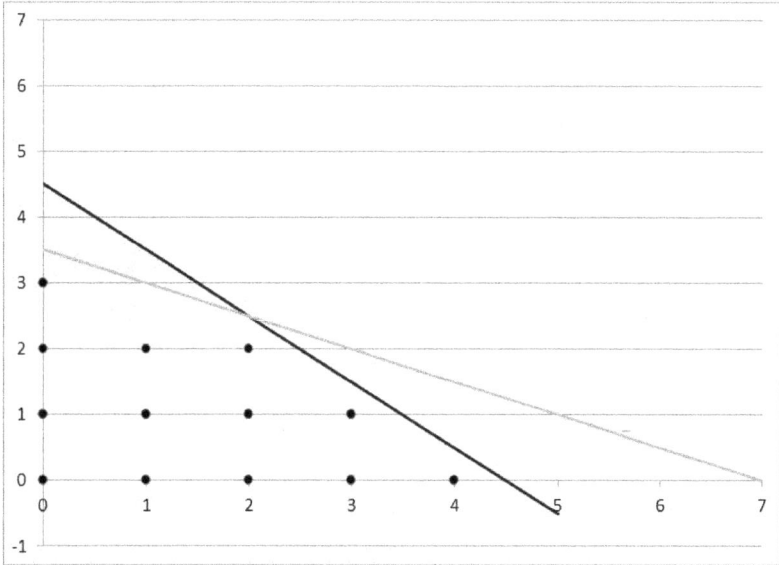

Figure 12 Integer programming in the feasible region.

Now we apply the same approach with the objective function, but only choose from solutions that have whole numbers. As shown below in Figure 13, by limiting our potential solutions to whole loads, the solution that maximizes our revenues occurs at (2, 2), or two loads each from Xavier and Yellow Springs. This combination gives us a maximum Total Net Revenues of $1400 ($300 x 2 loads from Xavier + $400 x 2 loads from Yellow Springs).

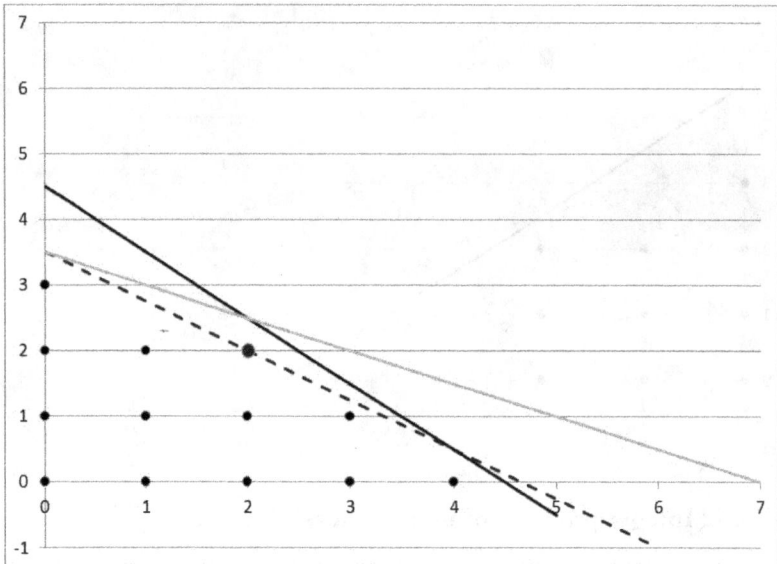

Figure 13 Whole number solutions applied to the objective function.

A minimization problem, such as minimizing the cost of a set of capital projects, is simply the mirror image of the maximization problem we just solved. Let's say our constraints were changed and we needed a minimum of 7 and 9 tons of phosphorus and nitrogen. Then our constraints would be greater than or equal to, instead of less than or equal to. This leads to a feasible region in grey, shown below in Figure 14. Now if our objective function was to minimize processing costs of $300 and $400 per load instead of profit, we simply take the same approach as before, but we start with high costs and slide the objective line down until we reach an extreme point solution. In this case, since all we did was make the problem a mirror, the minimum costs of $1600 would occur at the point (2, 2.5).

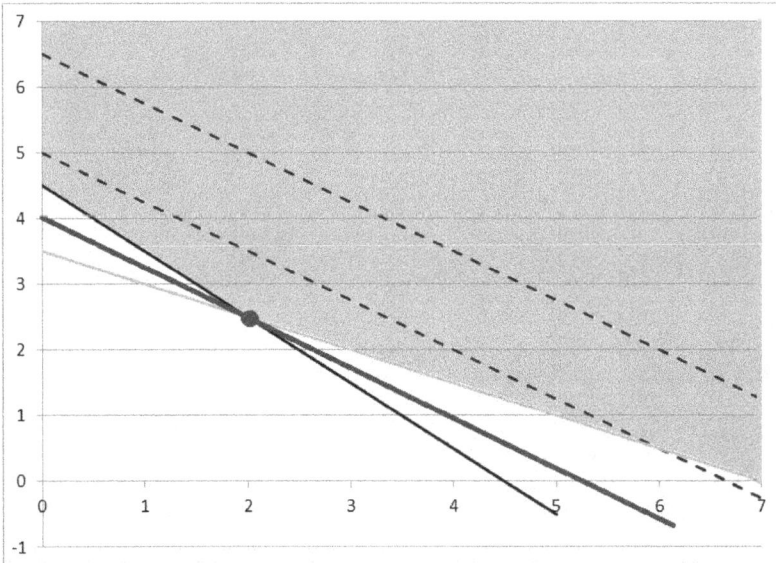

Figure 14 Minimization problem feasible region.

There are some special cases when setting up optimization problems that can cause difficulties in obtaining solutions. The two most important are when you have an unbounded solution or when you have an infeasible solution. An unbounded solution can occur when the constraints are minimums and the problem is maximization. For example, the graph in Figure 15 below shows a feasible region that is grey. If you keep sliding the objective function (dotted line) up to maximize the value, it will keep moving to infinity.

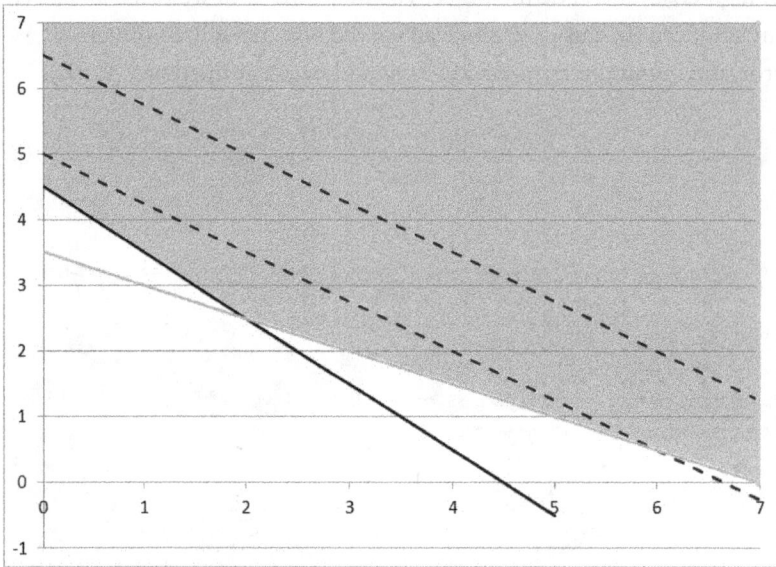

Figure 15 Unbounded solution.

An infeasible solution occurs when all of the constraints can't be met with one solution. In the example below in Figure 16, the top shaded area represents the feasible area for one constraint and the lower shaded area represents the other feasible area for a second constraint. Since there is no overlap, there can be no single solution given the existing constraints.

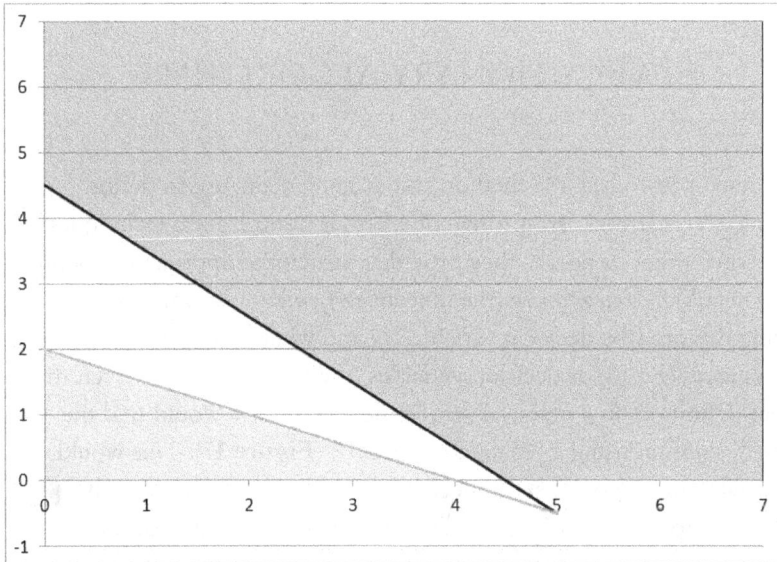

Figure 16 Infeasible solution.

5 EVOLUTIONARY ALGORITHMS

As discussed previously, the method systems engineers use to define problems is very useful, irrespective of whether optimization techniques are used to solve them or not, as they provide a structured approach to defining what we would like to achieve (the objective(s)), what our options for achieving this are (the decision variable(s)), and what limitations are placed on the objectives and/or decision variables (the constraint(s)). Given this problem definition, in a practical setting most engineers would find the "optimal" solution using a "manual" approach (**Figure 17**). This would involve selecting a "good" set of decision variables based on domain knowledge, experience and intuition and evaluating the utility of this solution in relation to the objectives and constraints, which would more often than not require the use of a simulation model. Based on this feedback, the engineer would try a different solution (i.e. a different set of decision variables) and evaluate its utility. This process would be repeated a number of times until the engineer believed they had found the "best" solution and had therefore "optimized" the problem.

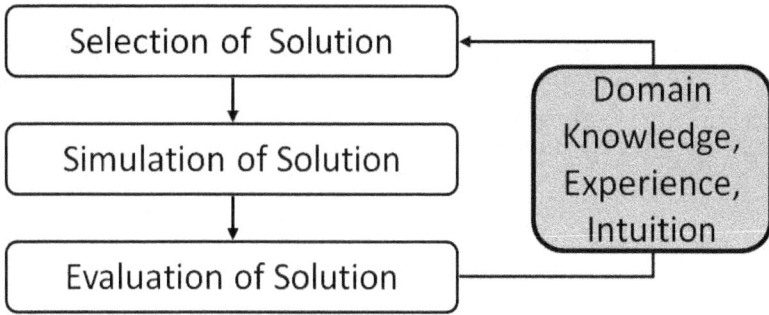

Figure 17 Manual simulation-based approach for finding optimal solutions.

Let's take the example of the three industrial plants discharging into the Shenandoah River introduced earlier and modify it so that the DEQ requirements correspond to maximum ambient nutrient levels in the river, rather than limits on the discharges. In this case, the objective is still to minimize treatment cost and the decision variables are still the treatment levels at each of the three plants, but the constraints are now the maximum allowable Phosphorus and Nitrogen concentrations in the river. So the approach to solving this problem would be to select treatment levels at each of the plants based on domain knowledge, intuition and experience, calculate the corresponding costs, and run a water quality model for the river to check whether the ambient water quality constraints have been violated or not. If the water quality constraints have been violated, the engineer would try increased treatment levels, calculate the corresponding costs, and run the water quality model again to check if the constraints are now satisfied. If the constraints are satisfied, treatment levels might be relaxed slightly in order to reduce costs while still satisfying the constraints, and so on.

Many engineers like this approach as it eliminates many of the concerns about optimization discussed earlier, such as:

- The "black box" nature of optimization methods since this approach is transparent and uses simulation models engineers are familiar with.

- Trust and credibility issues associated with the results produced by optimization methods, as the utility of solutions is assessed using known simulation models.
- The need for the development of special mathematical formulations and the use of techniques such as calculus and linear algebra.
- The need to simplify problem representations due to the inability of optimization approaches to handle Boolean, non-linear, or discrete integer constraints, for example, as the required degree of complexity is captured by the simulation model.

So wouldn't it be great if optimization approaches existed that ticked all of the above boxes! Well, such approaches do exist and are generally referred to as evolutionary algorithms (EAs). These methods work in exactly the same way an engineer would in that they select a potential solution, evaluate the utility of the solution with the aid of one or more simulation models, and then use this feedback to decide which solution to try next. The only differences are that (i) the decisions as to which options to try are made with the aid of evolutionary operators, rather than human judgement, intuition and experience, and (ii) the number of solutions considered as part of the trial-and-error process is much larger (Figure 18).

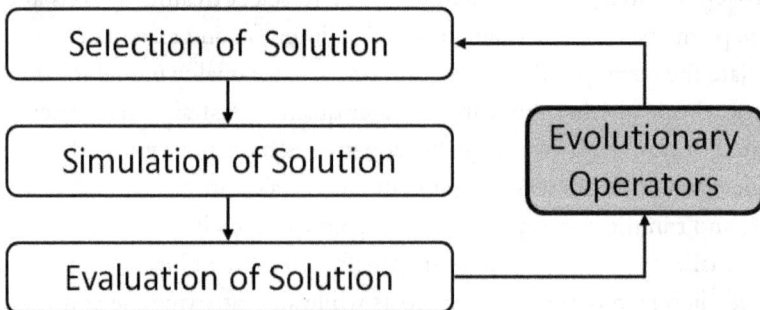

Figure 18 Process for finding optimal solutions when evolutionary algorithms are used.

So how do these algorithms figure out which solutions to try next? Now this is where it gets really interesting! The way these algorithms work is inspired by examples from nature, such as genetic evolution and the way ants forage for food. While we generally don't think of these phenomena as optimization methods, they actually are and are extremely efficient. For example, in evolution the genetic operators of selection (survival of the fittest), crossover, and mutation are very efficient at evolving offspring (solutions) that are better adapted to their environment than their parents. In another example, ants search for food using pheromone to identify the shortest path between their nest and a food source. So how can these principles be used to find optimal infrastructure solutions? In order to answer this question, let's look at a simplified example of the optimal design of water distribution systems (WDSs).

WDSs consist of the network of pipes, pumps, tanks and valves that is used for transporting water from water treatment plants to individual households. When designing such networks, the aim is to minimize overall system cost, while ensuring that each consumer receives sufficient water at an acceptable pressure. By using smaller pipes, costs are reduced. However, smaller pipes also increase pressure loss along the length of the pipeline, and may thus result in undesirably low pressures. Consequently, in this simplified example the aim is to identify the cheapest combination of pipes that satisfies pressure constraints. Let's have a look at how EAs that are based on genetic operators and the foraging behavior of ants can be used to solve this problem.

Optimization algorithms that identify better solutions with the aid of genetic operators are called genetic algorithms (GAs). When GAs are used for optimization, the decision variables (e.g. diameters for each pipe in the WDS) are represented by a string of numbers (equivalent to chromosomes in the genetics analogy), where each number represents the diameter of a particular pipe (Figure 19). Therefore, each chromosome represents a particular WDS design. In contrast to the "manual" optimization process outlined above, the GA procedure starts with the generation of a number (population) of different solutions (designs), rather than a single solution. These initial designs can be obtained based on experience, domain knowledge or intuition, or generated randomly, although the latter option generally increases the number of required iterations before an optimal solution is identified. Next, the "fitness" of each of the solutions in the population is assessed by calculating objective function values (e.g. the cost) and checking the constraints (e.g. whether all minimum pressure constraints are satisfied). As GAs are unable to cater to constraints explicitly, any constraint violations are generally converted to a penalty cost. Consequently, for the WDS design example considered here, the fitness function would consist of the physical cost of the system, as well as penalty costs for violating any of the pressure constraints. Whether a particular design meets the desired pressure constraints is determined using a hydraulic simulation model of the system under consideration.

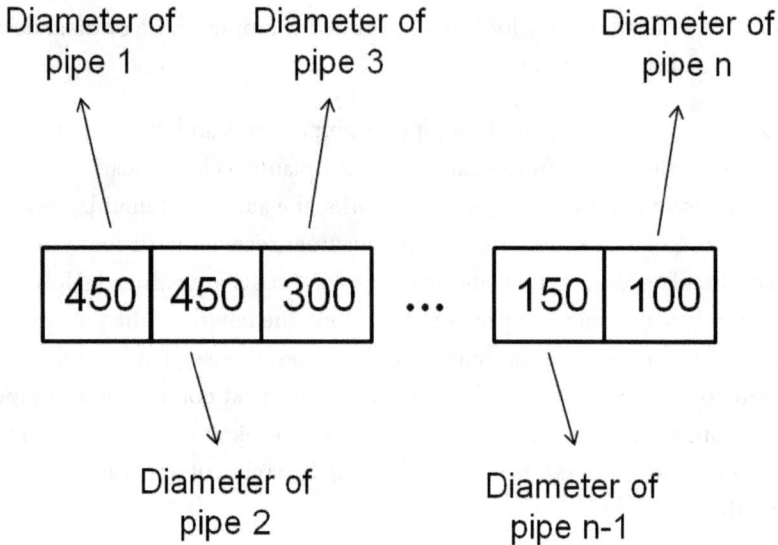

Figure 19 Representation of solutions (selected pipe diameters) for the simplified water distribution system optimization problem in the form of a chromosome (string of numbers). Each box ("gene") represents the selected diameter for one of the n pipes in the system, shown in terms of mm in the figure for the sake of illustration.

During the next step in the process, the fittest members of the population are selected for reproduction (Figure 20). This *selection* process generally consists of "tournaments" between randomly chosen pairs of chromosomes, where the fitter chromosome wins and the weaker chromosome is eliminated. This is similar to the concept of "survival of the fittest" found in natural evolutionary processes and enables the algorithm to move the search towards more promising regions of the solution space. The winning, fitter chromosomes have the opportunity to "reproduce", which involves the exchange of "genetic material" between randomly chosen pairs of "winning" chromosomes (the "parents") in a process called *cross-over* (Figure 20). As part of this exchange, each pair of strings is "cut" at a random location and the information after the cut is exchanged between chromosomes to form the "children" chromosomes, which contain genetic material from both parents (Figure 21). This has the purpose of gravitating towards better regions of the solution space. Finally, there is a small probability of mutation, where a small number of pipe diameters are changed at random (Figure 22). This serves the purpose of moving the search to different regions of the solution space in order to facilitate greater exploration and prevent stagnation and convergence to sub-optimal solutions. The resulting new generation of solutions is then subjected to the genetic operators of selection, crossover, and mutation, producing a new set of "children" solutions and so on. In this way, fitter and fitter solutions are evolved over the generations until optimal or near-optimal solutions are identified.

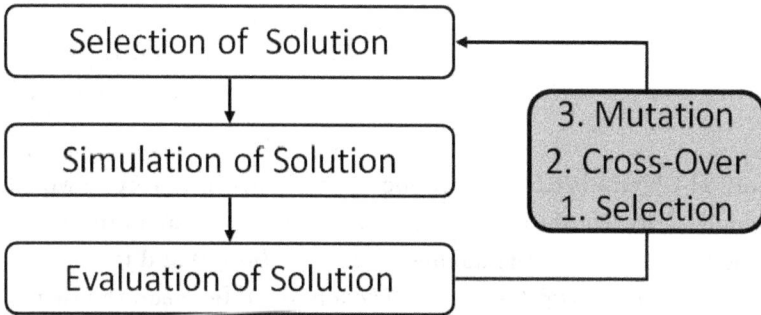

Figure 20 Process for finding optimal solutions when genetic algorithms are used, highlighting the genetic operators used to identify better solutions from one generation to the next.

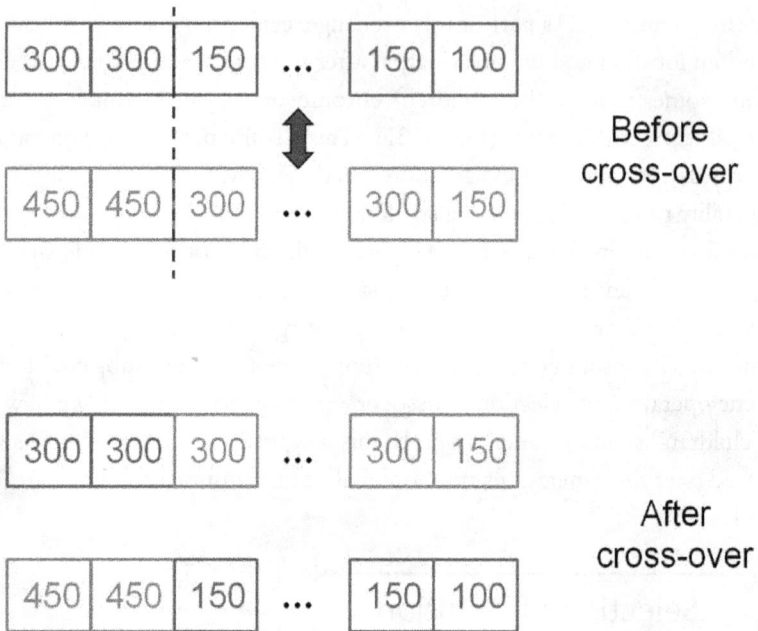

Before
cross-over

After
cross-over

Figure 21 Illustration of the genetic operator of cross-over. Two randomly selected chromosomes (solutions) are cut at a random location (indicated by the dashed line in the figure) and the information to the right of the cut is exchanged between the two chromosomes.

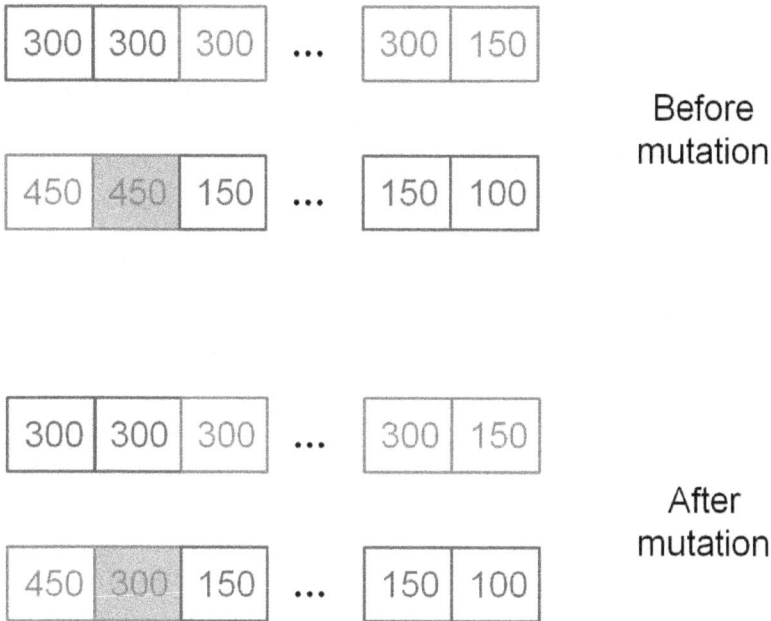

Before
mutation

After
mutation

Figure 22 Illustration of the genetic operator of mutation. Pipe diameters to be "mutated" are selected based on a very small probability (i.e. the vast majority of diameters are not affected by mutation) and then changed to another diameter at random, as illustrated for the pipe diameter shaded in the figure.

Algorithms based on the foraging behavior of ants are called Ant Colony Optimization Algorithms (ACOAs). ACOAs improve the quality of solutions from one generation to the next by using the principle of positive reinforcement. Ants are almost completely blind, yet they are able to find the shortest distance between their nest and a food source. This is achieved using an indirect form of communication. As ants forage for food they deposit a chemical substance called pheromone on the path they travel. Other ants are more likely to follow paths that have a higher concentration of pheromone. How this leads to the determination of the shortest distance from a nest to a food source can best be explained by means of an example. In the situation depicted in Figure 23a, a number of ants are travelling along the shortest route between their home (H) and a food source (F). If this shortest path is blocked by an obstacle (AB) (Figure 23b), such that there is now a longer path (HAF) and a shorter path (HBF) between the home and the food source, the ants will quickly determine which of the two paths is shorter. Initially, an equal number of ants will travel along the longer and the shorter path as there is no pheromone on either. However, as ants that travel along the shorter path from the nest to the food source reach the food source more quickly, and ants that travel from the food source to the nest along the shorter route will return home more quickly, more ants travel along the shorter route per unit time (Figure 23c). Consequently, more pheromone is deposited on the shorter path making it more likely to be chosen by the ants, further reinforcing the shorter route with pheromone. In addition, the pheromone on the path that is used less often evaporates, ensuring that almost all ants will travel along the shorter route (Figure 23d).

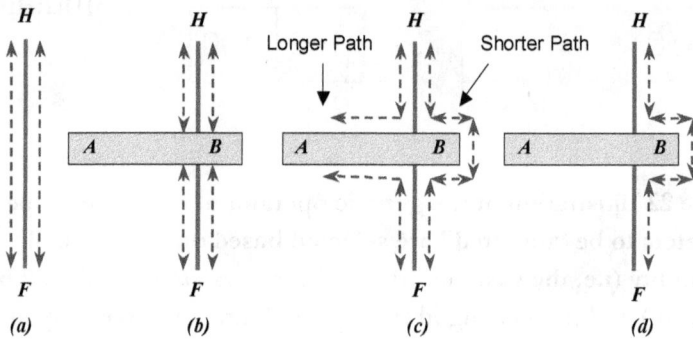

Figure 23 Illustration of the evolution of ant pheromone trails.

In order to apply ACOAs to infrastructure problems, such as the simplified WDS optimization example, each decision variable (e.g. pipe to be sized) has to be represented by a decision point which is equivalent to a junction at which a particular path has to be selected by an ant (Figure 24). At each decision point a number of paths are available, representing the different values a decision variable can take (e.g. pipe diameters). After paths have been selected at each decision point, a trial solution to the problem has been constructed (i.e. a complete WDS design has been obtained). Usually one trial solution is constructed by one artificial ant, and a population of solutions is generated by a colony of ants. As is the case when a GA is used, the fitness of a trial solution is calculated based on the physical cost of the pipes and a penalty cost associated with the degree of violation of the pressure constraints. Next, "paths" (e.g. pipe diameters) that have resulted in fitter solutions are awarded more pheromone than those that have produced weaker solutions, thus encouraging their choice in the next generation. Again, as is the case with GAs, better solutions are evolved from one iteration/generation to the next, until the optimal or near-optimal solution has been identified.

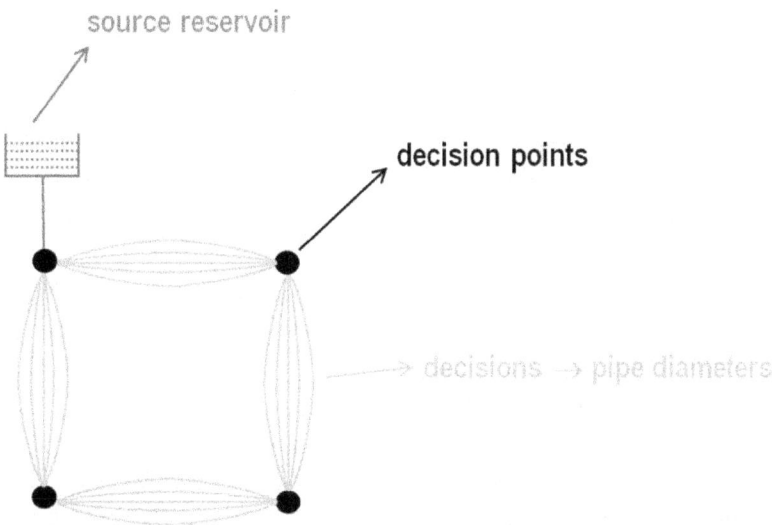

Figure 24 Illustration of the representation of the simplified WDS design problem for use with ant colony optimization for a very simple looped system with four pipes and a single source reservoir.

So if these algorithms work similarly to the way an engineer would, why do we need them? Comparisons between results from "manual" and EA-based optimization methods have invariably found that the latter can result in improvements in the order of 10%-50%. This highlights that the use of formal optimization methods can result in significant savings in costs and other resources, especially when dealing with large infrastructure projects, as well as increase accountability and transparency in an environment where resources (both financial and natural) are becoming increasingly constrained. In other words, the use of EAs enables solutions to be found that give you the "best bang for your buck", either in the sense of achieving a desired outcome with the smallest amount of resources or achieving the best outcome possible, given the available resources.

So why is the "manual" simulation-based optimization approach unable to do as well as EAs, especially since they make use of the same simulation model and essentially adopt the same approach to solving the problem? The main reason for this is that in many instances, the number of potential solutions (i.e. the number of potential combinations of decision variables) is incredibly large. For example, if there are 100 pipes that we would like to size in our WDS design problem and there are 6 discrete pipe options for each of these, the total number of potential solutions is $6^{100} = 6.53 \times 10^{77}$. Obviously, many of these combinations do not make sense from a practical perspective (e.g. going from the largest diameter in one pipe to the smallest in the next) and would not be considered as options by an experienced designer, but even if we can eliminate 99.99% of these solutions, the number of options remaining is still 6.53×10^{73}, which is incredibly large! This highlights the fact that we are only likely to explore a tiny fraction of the solution space when a "manual" simulation-based optimization approach is used, and which part of the solution space we explore is conditional based on the domain knowledge and experience of the engineer. The use of EAs enables us to explore larger regions of the solution space, as well as areas of the solution space we might not be familiar with (i.e. areas that go beyond the experience of the engineer), therefore enabling innovative solutions to be identified.

So, if we need to better explore the solution space in order to find the best and most innovative solutions, why don't we just evaluate all of the options and be done with it? Surely we can set up a computer to churn through the options without the need for fancy evolutionary operators? The answer to this question is simply one of computational efficiency. For example, if we would like to evaluate the utility of the 6.53×10^{73} solutions from our pipe sizing problem above and we assume that we could run the hydraulic simulation model required to check whether minimum pressure constraints have been satisfied 1000 times per second (which corresponds to a very computationally efficient model), it would take $\sim 2 \times 10^{63}$ years! So this would simply not be possible. This demonstrates that even though we try many more solutions when we use EAs (on the order of 50,000 to 100,000) than when we use the "manual" approach (on the order of 10-100), we still only evaluate a tiny fraction of the solution space. The fact that EAs are able to find globally, or near globally, optimal solutions for large problems while only evaluating a very small number of solutions is testament to the power and effectiveness of the evolutionary search mechanisms. EAs explore solution spaces in a very efficient manner because they not only have searching mechanisms that explore the search space widely, but also exploit information about the quality of the solutions considered in order to converge to global or near-global optima. These searching mechanisms have been honed over millions of years in nature to the point where they are now extremely efficient.

As EAs are essentially guided search methods, there is no guarantee that the globally optimal (i.e. "the" best) solution will be found, especially for large search spaces. However, they generally come close (although we don't really know what the best solution is for very large problems) and perform significantly better than the "manual" simulation-based optimization approach, as mentioned above. In contrast, some of the more mathematical optimization methods, like linear programming, are guaranteed to find the globally optimal solution. However, for most realistic problems, these methods can only be used once the problem has been converted to a simplified mathematical formulation. Consequently, the philosophies that underpin EAs and more traditional, mathematical optimization methods are very different in that EAs attempt to find near-optimal solutions to realistic problems (making use of existing simulation models), whereas the more traditional approaches find the optimal solution to simplified problems (e.g. they might solve a linearized form of the problem as they cannot utilize simulation models).

Given that in a practical setting, finding the "true" global optimum is less important than solving the actual problem, EAs are generally better suited to real-life applications than more traditional methods. Due to the loose coupling between the EA optimization engine, which decides which parts of the solution space to explore, and the simulation model, which evaluates how well the selected solutions perform in relation to the objectives and/or whether constraints have been violated (Figure 25), EAs can deal with discontinuities and non-linearities with ease, as long as these have been captured appropriately in the simulation model. In other words, if a problem can be simulated, it can also be optimized with the aid of EAs, as the EA essentially "bolts onto" the simulation model. This means that EAs can pretty much be applied to any problem in any problem domain, making them an extremely flexible and powerful optimization tool.

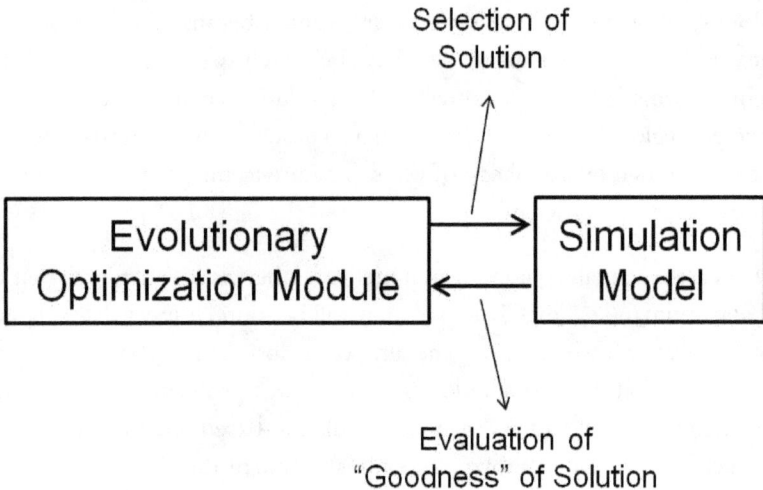

Selection of
Solution

| Evolutionary Optimization Module | → ← | Simulation Model |

Evaluation of
"Goodness" of Solution

Figure 25 Illustration of loose coupling between the evolutionary optimization module and the simulation model, where the optimization module can be "bolted onto" any simulation model. The optimization module identifies which solutions to try and passes these to the simulation model, which evaluates the utility of these solutions. This information is passed back to the optimization module, where it is used to determine which solutions to try next with the aid of evolutionary operators.

Another attractive feature of EAs is that they are not necessarily prescriptive in the sense of suggesting "the" optimal solution. This is because they work with populations of solutions and therefore produce a number of near-optimal solutions, which might be similar in objective function space, but quite different in solution space. This enables consideration of factors other than those captured in the mathematical formulation of the optimization problem when selecting the final "optimal" solution. It gives greater control to engineers in terms of using their judgement and intuition to select the final solution based on a number of good solutions "suggested" by the optimization algorithm. In this way, the EA is used to assist with "sifting through" the very large solution spaces that are a feature of real infrastructure problems in order to identify a set of near-optimal candidate solutions that can then be scrutinized by engineers to identify those that make most sense.

Another advantage of EAs is that they can deal with discrete decision variables. If we take the WDS optimization problem, when traditional optimization methods are used, the optimal pipe diameters will be given in terms of real numbers. However, in practice there are discrete commercially available pipe sizes. Consequently, if the optimal pipe diameter obtained using a traditional optimization method turns out to be 373mm, this will have to be rounded up to the nearest commercially available diameter. This will probably mean that the resulting solution is no longer optimal and is an example of the mathematically optimal solution that has been identified is meaningless in a real-life, practical context. In contrast, EAs can work with discrete and continuous decision variables. So in the WDS example, the decision variables would only consist of commercially available pipe diameters, as would the near-optimal solutions produced at the end of the optimization process.

Finally, yet another advantage of EAs is that they are able to deal with multi-objective problems with ease. For example, when designing WDSs, cost minimization is often not the only objective. In some instances, we would also like to maximize system reliability or minimize environmental impact via, for example, greenhouse gas emissions. In these instances, there is no single optimal solution, but a front of solutions, referred to as the Pareto Front, that represents the optimal trade-offs between the competing objectives (Figure 26). Due to the way in which optimal solutions evolve in EAs, they are able to generate the Pareto Front of optimal solutions in a single computer run, which is not the case for more traditional optimization approaches.

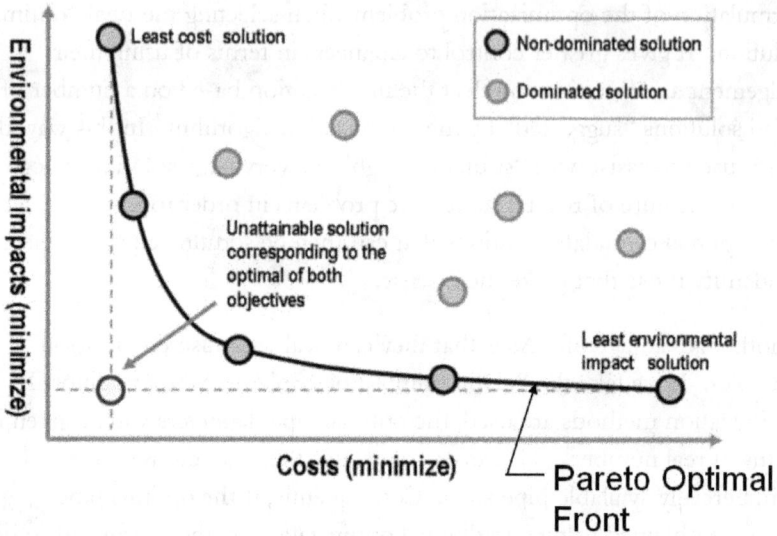

Figure 26 Illustration of the concept of Pareto optimality, where each of the dots represents a solution (design) that corresponds to different values of the objectives (the minimization of costs and environmental impacts in the example in the figure). Solutions that are on the Pareto Front provide the best possible trade-offs between objectives and are referred to as non-dominated solutions, whereas solutions that are not on the Pareto Front are referred to as dominated solutions.

Recommended References

If you would like to find out more about these techniques applied in similar context in more depth, we suggest the following papers:

Beh E.H.Y, Maier H.R. and Dandy G.C. (2015) Adaptive, Multi-Objective Optimal Sequencing Approach for Urban Water Supply Augmentation under Deep Uncertainty, Water Resources Research, 51(3), 1529-1551, DOI:10.1002/2014WR016254.

Foong W.K., Maier H.R. and Simpson A.R. (2008) Power plant maintenance scheduling using ant colony optimisation: an improved formulation. *Engineering Optimization*, 40(4), 309-329, DOI:

10.1080/03052150701775953.

Gibbs M.S., Dandy G.C. and Maier H.R. (2010) Calibration and optimization of the pumping and disinfection of a real water supply system. *Journal of Water Resources Planning and Management*, 136(4), 493-501, DOI: 10.1061/(ASCE)WR.1943-5452.0000060.

Maier H.R., Kapelan Z., Kasprzyk J., Kollat J., Matott L.S., Cunha M.C., Dandy G.C., Gibbs M.S., Keedwell E., Marchi A., Ostfeld A., Savic D., Solomatine D.P., Vrugt J.A., Zecchin A.C., Minsker B.S., Barbour E.J., Kuczera G., Pasha F., Castelletti A., Giuliani M. and Reed P.M. (2014) Evolutionary algorithms and other metaheuristics in water resources: Current status, research challenges and future directions. *Environmental Modelling and Software*, 62, 271-299, DOI: 10.1016/j.envsoft.2014.09.013.

Newman J., Dandy G.C. and Maier H.R. (2014) Multiobjective optimization of cluster-scale urban water systems investigating alternative water sources and level of decentralization. *Water Resources Research*, 50(10), 7915-7938, DOI:10.1002/2013WR015233.

Paton F.L., Maier H.R. and Dandy G.C. (2014) Including adaptation and mitigation responses to climate change in a multi-objective evolutionary algorithm framework for urban water supply systems incorporating GHG emissions. *Water Resources Research*, 50(8), 6285-6304, DOI:10.1002/2013WR015195.

Szemis J.M., Maier H.R. and Dandy G.C. (2014) An adaptive ant colony optimization framework for scheduling environmental flow management alternatives under varied environmental water availability conditions. *Water Resources Research*, 50(10), 7606-7625, DOI: 10.1002/2013WR015187.

Wu W., Maier H.R. and Simpson A.R. (2013) Multi-objective optimization of water distribution systems accounting for economic cost, hydraulic reliability and greenhouse gas emissions. *Water Resources Research*, 49(3), 1211-1225, doi:10.1002/wrcr.20120.

PART III: FOOD FOR THOUGHT

6 OPTIMIZING SMALL WATER DISTRIBUTION SYSTEM CIP

Now that we have seen how basic problems are set up, the next three chapters take a look at optimization applied to real world applications in the water resources utility sector. This chapter, Chapter 6, shows how optimization of a capital improvement plan (CIP) could be accomplished using a spreadsheet-based solver. Chapter 7 demonstrates how the processing power available in the cloud can help water, wastewater, and stormwater utilities take advantage of optimization. Chapter 8 shows how more complex decisions about tradeoffs in the triple bottom line (TBL) can be addressed in the interest of total water management.

Faced with both aging infrastructure and pressure from the public to keep water and sewer rates low, water and wastewater utilities are looking to make the most out of their capital budgets. Optimization can help utilities do the proverbial "more with less." An example follows of how real data from a small to medium sized water and sewer system could be used to optimize a CIP. Note that this small scale evaluation could be done using Solver in MS Excel. More complex projects would require more processing power provided by some of the more robust solvers and cloud-based computing.

This utility, whose data have been sanitized to protect privacy, collected information on the water distribution system, broken into discrete segments, which were rated by the following criteria:

- Pipe Size
- Pipe Material
- Break History
- Number of Complaints
- Adequacy of Fire flow
- Looping

These factors fed into a Deterioration Point Assignment Method, a scoring algorithm that awarded points for each criterion and an estimated cost to replace the pipe segment. Based on these factors, each pipe segment, or potential project, demonstrates greater need for replacement by scoring higher. A screen capture from the system that tracked the segments' points and costs is shown below in Figure 27.

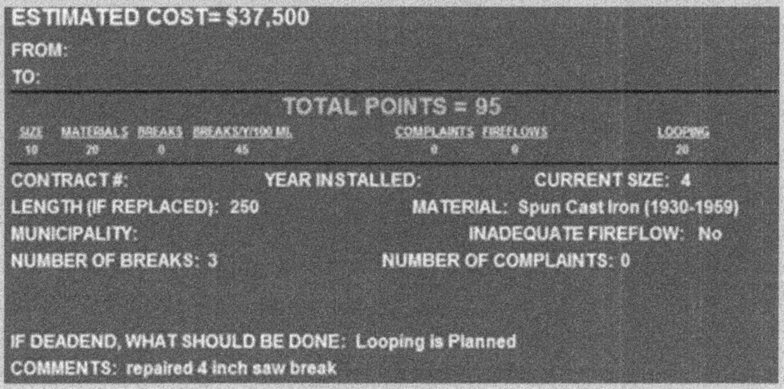

Figure 27 Screen capture of segment points and costs.

A straightforward approach to efficiently investing in capital improvement is to simply take as many projects from the top of the scoring list and put them into the CIP until the planned replacement budget is met. The system's sorting of projects using this approach (Top Down Score) is shown below in Figure 28:

LISTING OF CANDIDATES BY PRIORITY

STREET	SIZE	LENGTH	POINTS	COST
2" MAIN FROM: TO:	4	250	95	$37,500 VIEW DETAIL
2" MAIN FROM: TO:	4	919	85	$137,850 VIEW DETAIL
2" MAIN FROM: TO:	4	138	80	$20,700 VIEW DETAIL
2" MAIN FROM: TO:	2	58	75	$8,700 VIEW DETAIL
2" MAIN RD FROM: TO:	4	175	75	$26,250 VIEW DETAIL
2" MAIN FROM: TO:	2	104	75	$15,600 VIEW DETAIL
2" MAIN FROM: TO:	8	474	70	$71,100 VIEW DETAIL
2" MAIN FROM: TO:	4	701	60	$105,150 VIEW DETAIL

Figure 28 Top Down Score system.

Using the Top Down Score alternative, the CIP would consist of 12 projects for 845 points. Applying optimization to the score frameworks instead of the Top Down approach yielded substantial improvements. The optimization problem (a mixed integer linear program) is set up to maximize points (or risk) given a set budget of $600,000 and a potential problem set of 185 projects.

Maximize

TOTAL POINTS = Points001 x Project001 + Points002 x Project002 +... Points185 x Project185

Subject to:

Project Selection:

Project001 = 1 if you select, 0 if not
Project002 = 1 if you select, 0 if not
…
Project185 = 1 if you select, 0 if not

Cost Constraints:

TOTAL COST = Unit Cost001 x Project001 + Unit Cost002 x Project002 +… Unit Cost185 x Project185

TOTAL COST ≤ $600,000 (Budget)

Running the solver using the *Optimize Points* approach yields a CIP that has an 18% increase in total points, while saving 11% of the budget (Figure 29). This is due to the fact that the optimizer chose 7 less expensive projects with 45 points instead of 3 more expensive projects with 50 points.

Method	Projects	Cost	Points	Cost/Point
Top Down	12	$657,750	845	$778
Optimized	16	$585,450	1000	$585
Improvement	+ 4	11% Savings	18% Increase	25% Decrease

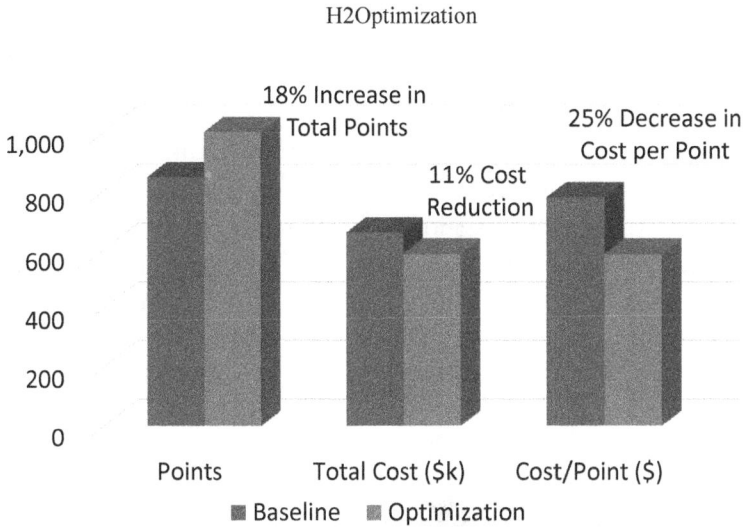

Figure 29 Expected performance improvement of optimization technique.

The utility wanted to look at a way to make the evaluation process more robust, so they developed an alternative based on a risk assessment framework. While risk is implied in the scoring approach, a more formal risk analysis was defined. Risk is defined as:

$$Risk = Threat * Vulnerability * Consequences$$

Where threat is the availability of the potential risk to occur. Since the risk of failure of a pipe is always possible, threat is defined to be 1. Vulnerability was calculated as a combination of:

- Probability of Failure (Expected Life).
- Function of Age, Material Type & Condition.
- Function of Break History and Complaints.

Consequences were estimated by evaluating potential

- Fatalities
- Serious Injuries
- Cost to Repair
- Lost Economic Impact to Community
- Loss of Confidence in Utility

When the same Top Down approach was applied to the utility's risk analysis, the recommended CIP consisted of 5 high risk projects being delivered. Applying the optimization approach to the risk-based framework results in even greater improvement than that of the scoring approach. The recommended CIP more than doubles the total risk reduction while still reducing expected cost by 8%, as shown in Figure 30.

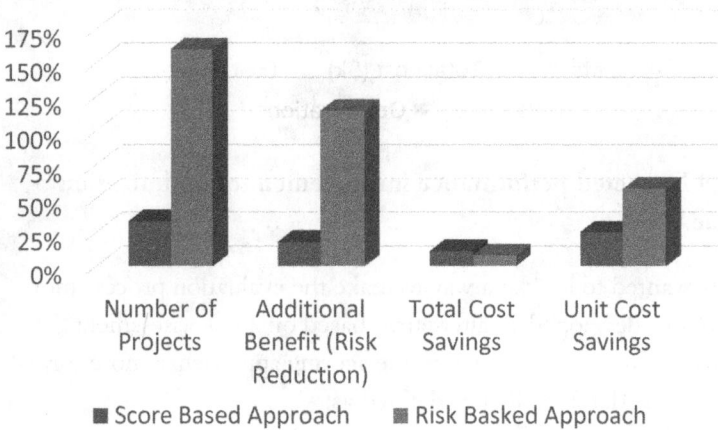

Figure 30 Expected performance improvement of risk vs. score techniques.

One of the advantages of having optimization and multiple analytical techniques is the ability to look at the recommendations under multiple scenarios before making a final decision. This is critical to real world problem solving to ensure that decision makers are not simply relying on a "plug and chug" mathematical algorithm. As you can see by the four analytical methods (Figure 31), different projects are identified by the various methods. However, projects ##28 and ##30 were included in the recommended CIP by all four methods and project ##91 was chosen by three. A planner could be very comfortable including these three projects in the CIP.

ID	Road	SCORE	SCORE - Opt	RISK	RISK - Opt
##28	H Lane	●	●	●	●
##30	P Street	●	●	●	●
##91	B Street	●	●		●
##26	D Court	●	●		
##83	G Brook Court	●	●		
##90	S Road	●	●		
##68	S Oak Drive	●	●		
##38	S Drive	●	●		
##69	B Avenue	●	●		
##63	W Avenue	●			
##76	M Street	●			
##91	W Drift Court	●			
##73	M Avenue		●		●
##05	B Lane		●		●
##52	B Drive		●		●
##50	P Street		●		
##78	P Park Drive		●		
##62	R Lynn Court		●		
##82	F Court		●		
##44	H Street			●	●
##25	B Street			●	●
##08	L Avenue, West			●	
##55	N Street				●
##13	B Street				●
##62	N Lane				●
##45	C Avenue				●
##75	L Avenue				●

Figure 31 Project selection by decision method.

Reference

Liner, B., Binning, D., and Gardner, N. (2009) "Risk Based Optimization of Pipe Rehabilitation and Replacement." *Virginia AWWA and Virginia Water Environment Association Joint Annual Meeting*, Sept. 13, 2009. Richmond, VA

7 CLOUD COMPUTING ENABLED OPTIMIZATION FOR WATER INFRASTRUCTURE

Optimization techniques can form the backbone of a formal decision support approach to help municipal water, wastewater and stormwater utilities identify lower cost, better performing plans and designs compared to current practice. Advances in cloud computing provide opportunities to take advantage of the exceptional computational power of multiple processors to crunch the numbers that enable evaluation of a vast number of alternatives, leading to the development of better solutions. These advances have made optimization techniques available to water, wastewater, and stormwater utilities that they would not otherwise have been able to use. When used effectively, an optimization approach can lead to better understanding of allowable improvement options and, ultimately, to lower-cost, defensible solutions.

An optimization model can extend the power of a utility's existing water distribution system, wastewater collection system or stormwater simulation. Utility and consultant modelers can shift their focus from tedious trial-and-error simulation runs to defining a range of improvement options and then letting the optimization search find the mix of options that yield improved system performance at lowest cost, or even consider other objectives.

Traditional simulation trial-and-error approaches to developing long term control plans generally rely on a hydraulic model to evaluate and modify a discrete set of solutions. Using an optimization process such as the one shown in Figure 32 lets planners and modelers solicit input from critical stakeholders (such as Operations, Engineering, Planning, and Management groups) to identify possible improvement options. Doing this enables the planners and modelers to identify the range of elements that could be included in a plan before being distracted by the constraints. Once the general framework is in place, optimization can be used to integrate the hydraulic model with financial and social concerns to develop and evaluate numerous low-cost, hydraulically superior solutions from which planners and managers can select with confidence.

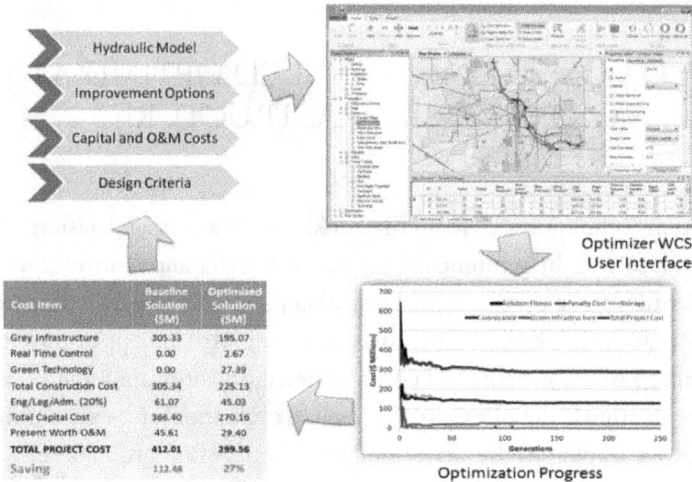

Figure 32 Optimization process (from Frey, et. al.).

The following three vignettes demonstrate how water and wastewater utilities are beginning to realize the potential of optimization in capital planning.

Sanitary Sewer Capital Planning in the US Pacific Northwest

A city in the Pacific Northwest region of the US experienced rapid growth and was forced to make sanitary sewer system improvements that minimized up-front capital investments, unfortunately leading to high ongoing O&M costs. The system is undersized in many areas resulting in dry weather overflows, surcharging manholes, and limitations on siting of new commercial and industrial businesses and residential development. Facing an estimated $130 million fee to upgrade the sewer system, the City aimed to not only reduce the upgrade price tag, but also to ensure public support. The City enlisted volunteers to join a citizen advisory group that met monthly to provide input and guidance to the technical engineering team.

During the course of the project more than 10 million individual trial plans and solutions were created and evaluated. Extensive sensitivity analyses tested population growth scenarios, model calibration accuracy, and water conservation. At the conclusion of the planning effort, the recommended plan was approved unanimously by the citizen advisory group and the optimization framework is estimated to save roughly $45 million over the life of the system.

Midwestern USA Combined Sewer Overflow Minimization

A Midwestern city has two service areas, one covering 40 km² (15 mi²) in combined sewer area and 120 km² (45 mi²) in separate sewer areas. Under current conditions, the city had approximately 60 overflows per year. The City entered into a Consent Decree aimed at reducing the number of overflows. In order to meet the demands of the decree, the City applied an genetic algorithm based optimization approach (like that discussed in Chapter 5) to its Master Plan Modeling Optimization Study. The project was designed to provide regulatory-compliant, low-cost Long Term Control Plan solutions for the City's combined sewer and sanitary sewer system areas. It was made more complex by the need for three different levels of service to be satisfied in terms of impact on the three rivers within the City's collection service areas. The objective was to meet the service levels and eliminate sewer overflows at lowest possible cost.

Alternatives considered in the analysis included new plants and facilities (tunnels, treatment plants, offline and inline storage facilities, relief sewers and pump stations), modification or expansion of existing infrastructure (treatment plants, storage facilities, pump stations, and pond improvements), sewer separation; sediment removal options, and infiltration and inflow (I&I) reduction practices.

The constraints were to meet the goal of no flooding from the sanitary sewer system when faced with a 5-yr 6-hr design storm event and standard hydraulic performance requirements with respect to minimum and maximum velocities in force mains, maximum hydraulic grade lines, and minimum freeboard requirements. Some of the most important demands due to the Consent Decree required models to develop solutions limiting combined sewer overflows (CSOs) to zero overflows at the primary outfall in a typical year. In addition, a maximum of 4 overflows in a typical year to two of the rivers and a maximum of 1 overflow in a typical year from CSOs tributary to the most environmentally sensitive river.

Sewer Basins in Southwestern US

A City decided to evaluate whether a mathematical optimization approach could help them reduce costs after its system-wide master plan led to recommendations that were too expensive to feasibly implement, yet the problems they faced could not go unaddressed. The objective of optimization was to evaluate feasible alternatives for two adjacent sewer basins and identify low cost solutions to contain the 5-year, 6- hour design storm event. Alternatives considered included inter-basin transfers, conveyance and wet weather flow control facilities (offline linear transport/storage facilities). The optimization was developed under two scenarios: a design including storage facilities, and a design without storage facilities. The original plan's estimated cost was over $100 million in capital costs. The optimized plans have estimated savings of $21 million (including storage facilities) and $20 million (without storage facilities, either way, roughly 20 percent capital cost savings could be achieved). In addition, in the optimized plans there is a significant reduction in total new sewer pipe length which provides additional benefits in terms of minimizing construction disturbances.

Reference

Frey, J., Hickman, T., Reust, W., Cronberg, A., Gonzalez, F., and Eubanks, S. (2015) "Large and Small Municipal Utilities Benefit as Optimized Decision Support Identifies Least-Cost Improvements for Wastewater and Stormwater Planning." Water Environment Federation Collection Systems Conference, Cincinnati, OH, April 2015.

8 TRIPLE BOTTOM LINE ANALYSIS

The complexity of the water supply planning process is exemplified by the need to incorporate the "triple bottom line" of sustainability - economic (financial), social, and environmental concerns - into integrated water supply plans. The constraints imposed on municipal water supply planners to comply with higher level (state and regional) policies, manage user demand, address uncertainty such as climate change, and focus on the triple bottom line, highlight the need for these planners to have optimization tools such as goal programming (GP) techniques at their disposal. Goal programming can provide solutions for multiple objectives at once, which could be beneficial for evaluating sustainability in integrated water resources management.

Historically, municipal water utilities either explicitly or inherently apply a greater weight to the financial aspect of sustainability. Figure 33 below shows how a decision maker may make use of optimization to help clarify the complex decision making required to balance the many aspects of sustainability.

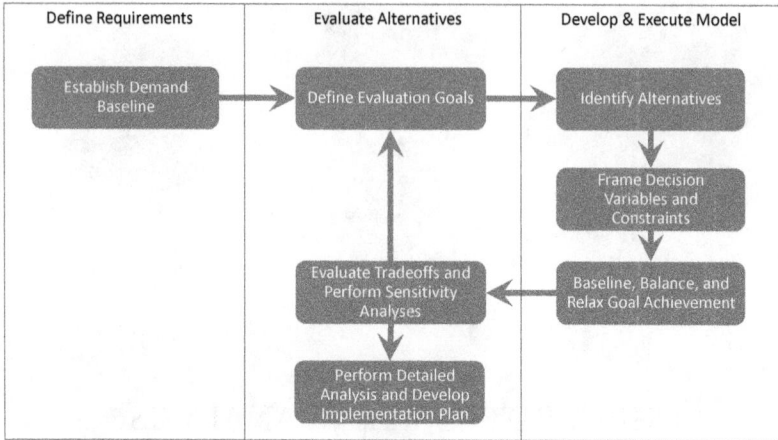

Figure 33 Complex decision making optimization.

The first step is to determine the requirements of the water resources management plan: What demand does the plan have to meet? What is the existing capacity? What demand management opportunities are available?

In order to demonstrate the feasibility of the use of the goal programming framework for municipal integrated water resources planning, real data and plans from a municipal water utility were entered into the model framework. Data were gathered from a utility that needed to increase its supply by 50% immediately, and triple the overall supply over the next 25 years. The costs and impacts on social and environmental goals were obtained for the following potential water supply alternatives: Seawater Desalination with Solar Electricity, Aquifer Recharge Using Lake Water with Pipeline to Infiltration Basin, Negotiated Exchange of Water Rights, New Wells, Recycled Water, and New Dam and Reservoir System.

For the demonstration, the goals were defined by the following performance measures:

Economic Goal – Lifecycle cost of Water Supply Alternatives. The lowest cost set of alternatives was normalized to 1.0, while the cost where water rates would cross the affordability threshold (2% of median household income) was set as the minimum economic goal achievement of 0.

Environmental Goal – Four measures make up the environmental goal: Water Quality and Environmental Issues (both set as categorical variables rated 1 (poor) to 5 (good)), volume of water recycled, and kW of electricity generated by the solar power plant.

Social Goal – With affordability integrated into the economic goal, categorical variables (1-5) for reliability, permitting, and institutional issues make up the social goal.

When baselining for lowest cost, the model's set of alternatives included bringing the dam online in year 1, new wells in year 8, recycled water in year 17, and a solar desalination plant in year 22. This combination yielded a 1.0 goal achievement value for Economic, 0.47 for Social and 0.31 for Environmental. When balancing all three goals, the set of alternatives included new wells, exchange of water rights, and recycled water brought into service as soon as possible and building a dam in year 20. The goal achievement for the three factors was 0.66 for Economic and Environmental and 0.8 for Social. The images in the Figures 34 and 35 below demonstrate these results.

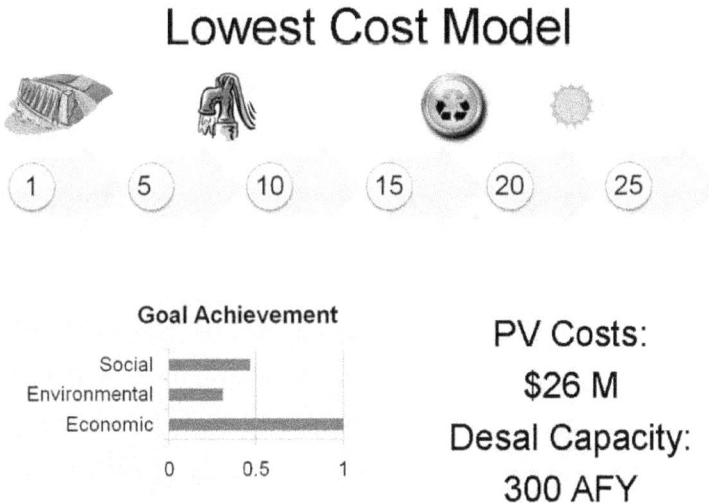

Figure 34 Example lowest cost solution of water supply alternatives.

Balancing the Goals

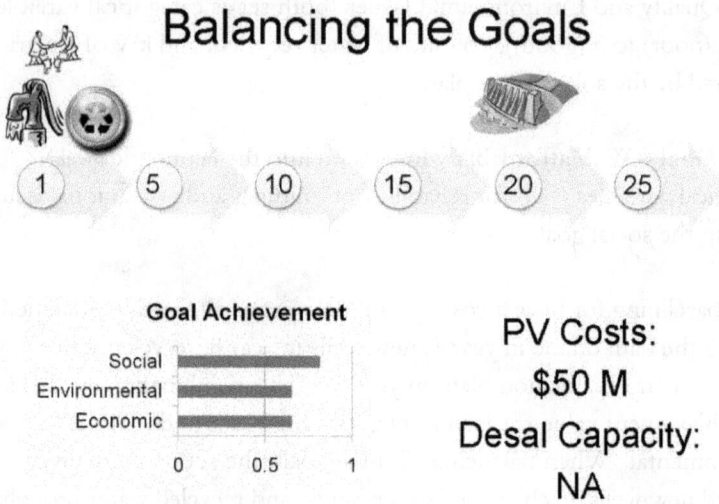

Figure 35 Example balanced solution set of water supply alternatives.

Trading Off Multiple Objectives

By performing these runs, the goal programming methodology can be used to develop tradeoffs between the various aspects of the TBL. The tradeoff curve in Figure 36 below details the relative costs for enhanced environmental and social goal achievement.

Tradeoff Cost vs. Goal Achievement

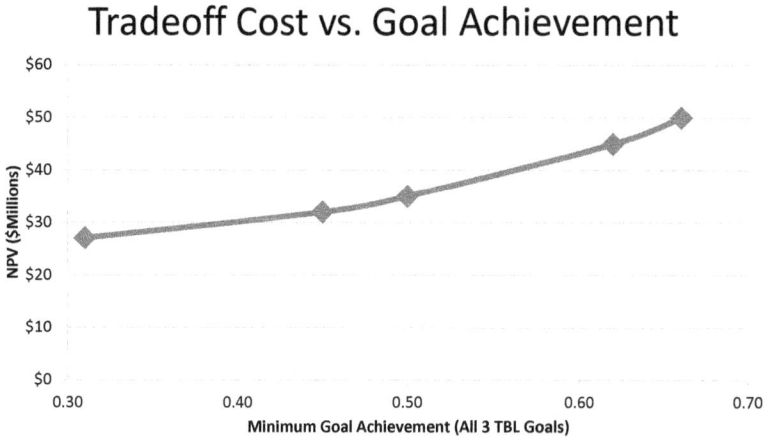

Figure 36 Tradeoff curve.

At the lowest cost ($26 million) the minimum goal achievement is 0.31. Looking at the shape of the curve, it appears that the slope changes around the 0.5 threshold, with a cost of $34 million. The decision makers could review this information and conclude that for a 31 percent cost premium ($34-$26 million) the utility could achieve over a 60 percent increase in environmental and social goal achievement (0.5-0.31). This type of analysis can be highly beneficial to utility planners in convincing the public, or even their own regulatory bodies, to incorporate the non-financial aspects of a plan into the decision process.

It is important to note that this model is intended to identify combinations of alternatives that can be evaluated at a more detailed level. By using goal programming enumeration of alternatives is minimized, which can be helpful when looking at a combination of multiple alternatives in different order over a 25-year planning horizon. In addition, visibility is provided into the relationship between the three components of the TBL.

From this tradeoff curve, the cost of including environmental and social factors into the analysis can be identified. In addition, the tradeoff can provide information to aid a decision maker in identifying the most politically and technically feasible set of alternatives - that is, how much environmental and social goal progress the decision maker can achieve at an acceptable cost.

The details on the set-up and implementation of the model, including how the goals were normalized to balance economic, environmental, and social components is shown in the following section. The final section in this chapter discusses some ideas on sustainability metrics which could be used to develop goals.

The Goal Programming Model and Implementation

Each of the potential alternatives is defined in terms of capacity, cost, and contribution to the performance measures. In some cases, the alternatives will have unit contribution toward goal achievement; while in others, the measure will be calculated after the aggregation of the impacts of the combination of selected alternatives.

Once the normalizing factors are determined, a maximin problem is solved to **Balance** the goal achievement by maximizing the minimum goal achievement for all three goals. The achievement function for the mixed integer, nonlinear program can be defined as:

Maximize $\quad Z = z$ $\hspace{4cm}$ (1)

Subject to:

$$z \leq \frac{\sum_{i=1}^{m} \sum_{j=1}^{n} c_{jEconomics} x_{ij}}{G_{Economics}} \qquad \text{Economic goal achievement} \qquad (2)$$

$$z \leq \frac{\sum_{i=1}^{m} \sum_{j=1}^{n} c_{jEnvironmental} x_{ij}}{G_{Environmental}} \qquad \text{Environmental goal achievement} \qquad (3)$$

$$z \leq \frac{\sum_{i=1}^{m} \sum_{j=1}^{n} c_{jSocial} x_{ij}}{G_{Social}} \qquad \text{Social goal achievement} \qquad (4)$$

Where z is the deviation variable for the goal achievement, G_k is the maximum goal achievement threshold for $k \in$ {Economic, Environmental, and Social}; x_{ij} are binary decision variables where i is the year and j is the alternative. For each of the m years in the planning horizon, if the model recommends that alternative j is brought into service in year i, then $x_{ij} = 1$, else $x_{ij} = 0$; $c_1, c_2, \ldots, c_n$ are contribution coefficients that represent the marginal contribution of the decision variable to the Economic, Environmental, and Social goals, and n is the number of alternatives (decision variables).

The primary constraints are related to capacity (supply must exceed demand, but not have wasted excess capacity) and uniqueness (an alternative can only be built once – for example, you can build a dam and reservoir one time, any upgrades would then be considered a new project). The constraints for these conditions, calculated for each year, are as follows:

Capacity Constraints

$$CAP^i_{Utility} = CAP^{i-1}_{Utility} - CAP^i_{Degradation} + \sum_{j=1}^{m} x_{ij} CAP_j$$

(Total Capacity including new alternatives and degradation of existing supply)   (5)

Where $CAP^i_{Utility}$ is the total water supply capacity of the utility at time i, CAP_j is the incremental capacity of alternative j, and $CAP^i_{Degradation}$ is any degradation of existing sources of supply, such as permitted flow reduction, plant retirement, or expiration of purchase agreements.

$CAP^i_{Utility} \geq D_i$ (Capacity Sufficient to Meet Demand)   (6)

$CAP^i_{Utility} \leq CAP_{Max}$ (Limitation of "overbuilding")   (7)

Uniqueness

$$\sum_{i=1}^{n} x_{ij} \leq 1 \text{ for each } j = 1..n \text{ alternatives} \qquad (8)$$

Where D_i is the demand in period i and CAP_{Max} is the maximum value of capacity to ensure the utility doesn't build excess capacity that it won't utilize during the planning horizon. The determination of CAP_{Max} is specific to each utility. There are a number of options to define the number (or set of numbers). The factor could be determined as a single number (for example, the demand at the end of the planning horizon), a percentage of capacity in each year (say, 150 percent of peak demand), or the demand a fixed time into the future (i.e., meet the demand 10 years from current year).

Goal Programming Model Execution

After defining the goals, the maximum goal achievement threshold G_k for $k \in \{Economic, Environmental, and Social\}$ is established in order to normalize the goals. This **Baseline** can be accomplished by maximizing each goal individually without any contribution from the other goals. Once the normalizing factors are determined, a maximin problem is solved to **Balance** the goal achievement by maximizing the minimum goal achievement for all three goals, as defined in equations 1 through 4.

The value of the objective function in the balanced run is the maximum minimum goal achievement for all three goals, or $Z_{Balance}$. After obtaining the maximum goal achievement threshold $Z_{Balance}$, the model can be rerun with the model objective to minimize cost (Maximize Economic Goal Achievement) subject to a **Relaxed** minimum goal achievement for the environmental and social goals set at intervals below $Z_{Balance}$

$$\text{Maximize } Z = \frac{\sum_{i=1}^{m} \sum_{j=1}^{n} c_{jEconomics} x_{ij}}{G_{Economics}} \qquad (9)$$

Subject to:

$$z' \le \frac{\sum\limits_{i=1}^{m} \sum\limits_{j=1}^{n} c_{jEnvironmental} x_{ij}}{G_{Environmental}} \quad \text{Environmental goal achievement} \quad (10)$$

$$z' \le \frac{\sum\limits_{i=1}^{m} \sum\limits_{j=1}^{n} c_{jSocial} x_{ij}}{G_{Social}} \quad \text{Social goal achievement} \quad (11)$$

Where z' is the relaxed minimum goal achievement level.

Potential Sustainability Measures

The following are some potential sustainability measures that could be used when making resource allocation decisions. The lists are obviously not exhaustive, and the relevance of these (or any measures) will vary depending on the local situation. These lists are merely to challenge your thought processes in defining the problem as we seek sustainable solutions for our water, energy, and other natural resources. For more information on sustainability reporting, we recommend Sustainability Reporting Statements for Wastewater Systems by The Water Environment Federation (2012) and the *Envision™ Sustainable Infrastructure Rating System* by the Institute for Sustainable Infrastructure (ISI).

Social Sustainability Measures
Usually considered the third pillar of the triple bottom line, social sustainability is by far the most difficult to quantify. One way to look at it is through a framework based on Maslow's hierarchy of needs. Figure 37 below shows how basic needs, like access to water, must be met first, before focusing on resilience and safety. Once those fundamental goals are met, then efforts related to involvement, enjoyment, and justice can be addressed.

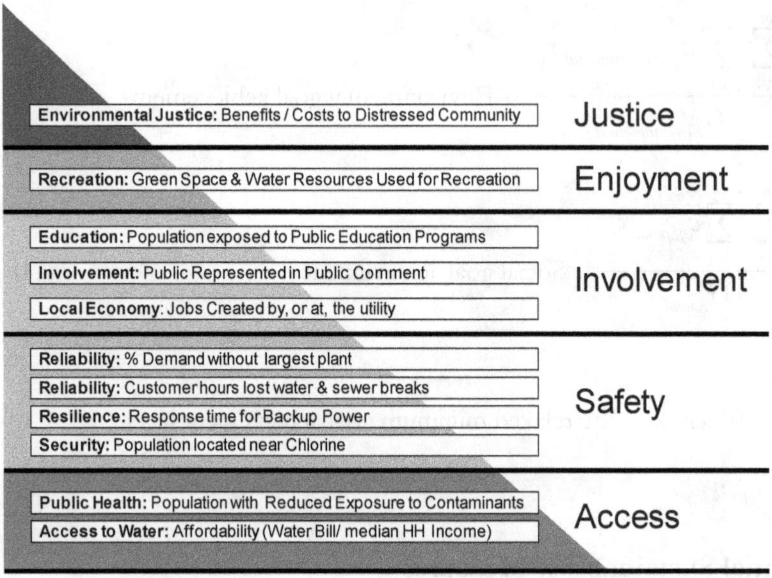

Environmental Justice: Benefits / Costs to Distressed Community	**Justice**
Recreation: Green Space & Water Resources Used for Recreation	**Enjoyment**
Education: Population exposed to Public Education Programs Involvement: Public Represented in Public Comment Local Economy: Jobs Created by, or at, the utility	**Involvement**
Reliability: % Demand without largest plant Reliability: Customer hours lost water & sewer breaks Resilience: Response time for Backup Power Security: Population located near Chlorine	**Safety**
Public Health: Population with Reduced Exposure to Contaminants Access to Water: Affordability (Water Bill/ median HH Income)	**Access**

Figure 37 Social sustainability measures in framework for basic needs (from Liner, deMonsabert, and Morley, 2012).

ACCESS

Average water bill / median household income – the basic affordability measure used by U.S. EPA, this is often calculated as the Hypothetical Water Bill for Non Discretionary Water Use / Median Household Income.

Population with reduced exposure to contaminants – a public health measure.

SAFETY

Percent of the minimum demand utility can supply when the largest water treatment plant is non-functional - A reliability of supply-focused indicator at the treatment plant level.

Expected duration to meet minimum demand on backup power after power loss – A resilience of operations indicator.

Population within a specified distance of chlorine storage – A physical security indicator related to the potential exposure to toxic chlorine leak.

Customer – Hours of service lost due to water main breaks - A reliability of supply-focused indicator at the distribution system level.

Customer – Hours of service lost due to sewer collapses - A reliability of supply-focused indicator at the sewer collection level.

Alternatives also include:
- Sewer collapses per km (or mile) of pipe,
- Water main breaks per km (or mile) of pipe
- Average time to repair main break (or sewer collapse)

INVOLVEMENT

Number of jobs created at utility by water supply project – local labor market impact of water system expansion, conservation program, or reuse system development.

Number of jobs created in community by water supply project - local labor market impact of industry brought to the area or made possible by water system expansion, conservation program, or reuse system development.

Number of people participating in educational programs provided by utility – a social outreach measure, public education programs are ubiquitous at water utilities. These programs include newsletters mailed with water bills, advertisements encouraging conservation, classroom visits by utility staff, or field trips to a treatment plant or education center. This measure exemplifies the challenges faced when developing social indicators. The "number of people contacted by educational programs" is measurable, which is a strength. However, an education or public outreach program may take years to achieve the intended results, if at all, therefore, measurement of the effectiveness of a program is challenging.

Percent of impacted population represented in public comment opportunities – a community involvement indicator.

Customer satisfaction – Customer satisfaction is also related to community involvement, but typically requires a survey to be directly measured, or indirectly calculated through proxy measures such as number of complaint phone calls received by the customer service center.

ENJOYMENT

Area created for parks and green space – a community outreach measure by providing parkland for the community.

Number of people using water resources for recreation - a community outreach measure by providing fishing, boating and other water based recreational opportunities for the area.

JUSTICE

Benefits to distressed communities / costs borne by distressed communities – an environmental justice measure to see if the beneficiaries of the expansion of the water system are proportionally bearing the cost or being provided the benefits.

Environmental Sustainability Measures

Water recycling - or wastewater reuse, provides an enhanced impact as recycling reduces both the withdrawals from raw water sources and the discharges to receiving bodies. This fact makes water recycling a key measure for environmental sustainability.

Water use – this category of measure encompasses consumption in total and in average by customer class. Customer classes may include residential, commercial, industrial, agricultural and governmental customers. The usage may be total water withdrawals, total water sales volume, total volume of wastewater discharged, average water usage by customer class, and average wastewater collection by customer class. A number of these indicators are useful in developing both demand forecasts and measuring the impact of demand management or conservation programs at the utility.

Impact of withdrawals – The watershed ecosystems can be affected with respect to quantity and quality by withdrawals.

Energy usage – Energy usage has many impacts from production and consumption. The basic measures of kWh used per volume of water or wastewater are the starting points. Energy costs per volume of water or wastewater are the next level of measures. These measures, plus the percentage of energy used from renewable sources (solar, wind, hydro, geothermal, tidal), are measures that can be used in decision making.

Total discharges – The total mass of nutrient load discharged by Biochemical Oxygen Demand (BOD), Suspended solids (SS), Nitrogen (N), and Phosphorus (P).

Biodiversity-rich habitats – The area of land managed that provides a habitat encouraging biodiversity. This measure is relevant when a new source (dam for a reservoir, for example) will reduce the habitat.

River quality – Specific measures may include flow, nutrients, and sediment loads.

Greenhouse Gas (GHG) emissions –The carbon footprint, including the identified GHGs: Carbon Dioxide (CO_2), Methane (CH_4), Hydrofluorocarbons (HFC), PFC, and CO_2 equivalents. In addition, it includes indirect emissions from energy consumption by suppliers (electric utility, steam generation, etc.) and the emissions from utility vehicles.

Economic Sustainability Measures
The most easily understood set of measures are the economic (financial) goals, as they relate to well documented issues such as budgets and rates:

Customer payments – this category of measure encompasses operating revenue (such as water sales, wastewater fees, reclaimed water sales, stormwater fees, impact fees, and miscellaneous fees) by customer class (residential, commercial, industrial, agricultural and governmental), or a combination of the types.

Operations & Maintenance (O&M) costs per unit volume – this category of measure focuses on O&M costs (typically labor, chemicals, materials, and energy). By looking at the unit costs per volume, the relative efficiency of the treatment and distribution processes can be assessed.

Payroll – describes the amount of money paid to the local economy through wages. This is an important measure when expansion creates new jobs in the community, and the marginal payroll increase could actually be a social benefit.

Interest on debt + dividend payments – debt management is important to any operations. Interest on debt reflects the result of the bond interest rate, which is a reflection of the management and stability of the utility.

Debt Ratio – can be expressed as total outstanding debt as a percentage of a regulatory debt ceiling, as Debt/Equity from the utility's balance sheet, or as a coverage ratio (Net Income divided by debt service payment). A debt ceiling may be regulated and the coverage ratio is generally part of the bond covenant when utilities issue revenue bonds. One of these measures could be useful in decision making when borrowing to fund capital projects is a major component.

Return on Assets (ROA) – defined as Net Income divided by total assets. Given the high value of assets, especially buried assets, that a utility has relative to the desire to keep water rates low, ROA tends to be lower at municipalities than in the private sector. Also, with the large volume of long term assets (for example, distribution pipes with 50 year lifetimes), any changes in water supply alternatives may have a negligible effect on ROA.

Asset renewal – focuses on the replacement of older assets with newer, more efficient assets. Since pipes are generally the largest asset class, this measure is probably better used for an asset management program or reporting than for water supply planning.

Annualized cost of water supply alternative – in order to address both O&M and capital costs, the annualized cost of an alternative is used. Generally, this measure by itself can serve as the Goal for the decision analysis.

Net Income – a key factor related to a number of the TBL classes of measures including Return on Assets, Customer Payments, Operating Ratio, and Debt Ratio.

Operating Ratio – a different manner of expressing Net Income, defined as Operating Costs divided by Operating Revenues (O&M plus debt service). A value of less than 1.0 implies self-sufficiency and a positive net income.

References

Liner, B. and deMonsabert, S. (2011) Balancing the Triple Bottom Line in Water Supply Planning for Utilities, *ASCE Journal of Water Resources Planning & Management*, 137 (4), 335-342.

Liner, B., deMonsabert, S., and Morley, K. (2012) Strengthening Social Metrics Within the Triple Bottom Line of Sustainable Water Resources. *World Review of Science, Technology and Sustainable Development*, 9 (1).

ABOUT THE AUTHORS

Dr. Barry Liner, P.E. is an industrial engineer with over 25 years' experience in the water resources field. He served as a professor in Civil, Environmental, and Infrastructure Engineering at George Mason University, where he taught courses in optimization modeling and engineering economics, infrastructure finance, engineering management, and sustainable development. Dr. Liner also served as Director of International Engineering Programs at Mason, leading water, sanitation and hygiene projects in rural Amazonian and Andean Peru. Throughout his career he has worked for organizations such as Black & Veatch, The World Bank, and he served as Director of the Water Science & Engineering Center at The Water Environment Federation. In his work at Applied Engineering Management Corporation and Aldera, he leads business process improvement efforts as a certified CMMI Lead Appraiser. He holds a Ph.D. in sustainable water resources and an M.S. in urban systems engineering from George Mason University, a B.A. in Economics from Virginia Tech, and a graduate certificate in business administration from Georgetown University. Barry resides in Virginia, USA.

Dr. Holger Maier is a professor at the University of Adelaide's School of Civil, Environmental and Mining Engineering. Dr Maier has won numerous awards for his contributions to modeling and decision analysis in the water resources field. He earned his Ph.D in Environmental/Water Resources Engineering and B.E. in Civil Engineering from The University of Adelaide. Holger resides in Adelaide, South Australia.